新工科建设之路·计算机类新形态立体化教材

计算机网络专业英语

许 勇 冯 锦 编著

电子工业出版社
Publishing House of Electronics Industry
北京·BEIJING

图书在版编目（CIP）数据

计算机网络专业英语 / 许勇，冯锦编著. -- 北京：
电子工业出版社，2025. 1. -- ISBN 978-7-121-49148-1

Ⅰ. TP393

中国国家版本馆 CIP 数据核字第 2024XJ5940 号

责任编辑：邱瑞瑾
印　　刷：三河市龙林印务有限公司
装　　订：三河市龙林印务有限公司
出版发行：电子工业出版社
　　　　　北京市海淀区万寿路 173 信箱　　邮编　100036
开　　本：787×1092　1/16　印张：12　字数：368.64 千字
版　　次：2025 年 1 月第 1 版
印　　次：2025 年 1 月第 1 次印刷
定　　价：39.00 元

凡所购买电子工业出版社图书有缺损问题，请向购买书店调换。若书店售缺，请与本社发行部联系，
联系及邮购电话：（010）88254888，88258888。
质量投诉请发邮件至 zlts@phei.com.cn，盗版侵权举报请发邮件至 dbqq@phei.com.cn。
本书咨询联系方式：（010）88254173 或 qiurj@phei.com.cn。

前　　言

专业英语教学没有统一的标准，也很难制定统一的标准，还面对具体而现实的难题，即师资类别的选择。同样的问题自然地延伸到了教材的编著和教辅材料的制作中。英语课和专业课的教学模式大相径庭，各有侧重，各有利弊。因为行业的差异，相同词汇在不同专业领域会有不同的含义。本教材的编著者在多年专业英语的教学实践中发现，由于目标不明确、手段不匹配，教师在教学中常常无所适从，学生也难以高效地构建知识体系和系统地提升技能水平。要实现专业课和英语课齐头并进，就不能放弃目标和方向，要清醒地认识到实际操作的教学手段还需要经过长期艰难的实践和探索。

就专业英语而言，其目标是学好英语。理想的专业英语教学是在学习者已具备一定专业知识的情况下，以该专业英文资料为载体来系统地提升英语语言能力，包括掌握科技英语的特点，积累专业词汇和术语，使科技英语的听、说、读、写和翻译能力得到有针对性的提升，培养以英语为工具来拓展和深化专业知识的能力。更高的目标是养成使用英语的习惯，用另一种语言培养新思维，开启另一个看待问题的新角度。不同的语系具有语义表达能力的互补性，由此带来了思维模式和文化习俗的微妙差异，即多一门语言、多一个世界。

直观地看，专业英语的教学是英语教师的职责，但实际情况是英语教师不具备强大的专业背景。语言的作用是承载和传播思想及信息，语言工作者必须知道自己在说什么，如果对想表达的对象不甚了解，那么再华丽的语言也是纸上谈兵。而专业课教师一般不会系统地关注语言本身的特点，不具有语言教学经验，因此很难对学生的语言能力进行专项拓展和训练，往往把专业英语课教授成了双语专业课。如果只是实现了专业词汇和术语的线性积累，而没有达到举一反三的效果，那么专业英语课就失去了其本身的必要性，以及其应有的课程价值和地位。

纵观目前专业英语教材的编写、编著情况，专业课教师基本按专业课体系对英文资料进行编排，力求覆盖完整的专业领域；而英语教师则按语法体系进行整合，重点讲解英语的语言特征和规律。前者编写、编著的专业英语教材以专业课为载体来学习英语，通过一种一一对应、线性积累的方式提升专业英语水平，更适合已掌握相当专业知识的学生和教师；而后者编写、编著的专业英语教材用语言深化专业知识，更适合英语掌握得比较好的学生和教师。当然，也有对上述两种方式进行机械拼接的，但语言和专业泾渭分明，存在明显的裂痕。

专业知识和语言能力既能相辅相成，互为目的和手段，也能"相爱相杀"，顾此失彼，导致不同结果的分水岭就在于对教学元素的空间粒度，以及专业知识和语言能力结合时机的准确把握。此外，也可以另辟蹊径，充分利用以先进移动通信平台为主导的现代化信息存储与交互手段，充分激发学习者的学习动力，适应当前学习者的生活和学习习惯。

本教材的教学设计基本策略如下：

多种形态，深化融合；难易搭配，动静结合。

角色互换，内化需求；理性主导，提升语感。

本教材改变教材的传统属性，使教材和对应的教学过程采用立体化的新形态，以线下训练为主、线上学习为辅，通过多种承载媒体、多种学习渠道完成教学材料和教学过程同步设计和制作，全面记录学习者的学习过程，并作为形成性考核的一种客观指标，实现线上和线下、专业和语言、学习和应用的有机结合。本教材以非线性的迭代方式灵活编排，颗粒化、立体化地呈现教学内容，以实现时空毗邻、多维度融合和关联，促进语言规律的实时总结和及时应用。就本教材教学元素而言，无论是专业知识点，还是语言点，适度即好、及时为佳。本教材素材的选择力求覆盖专业学科的全貌，并保留了传统教材的基本专业知识体量，但在篇幅上不强求整齐划一，以牺牲知识体系的局部完整性满足解决问题的必要性、可能性和高效性。本教材的前半部分设计为专业容易而语言复杂，侧重于对专业英语特点和规律的理解和掌握；后半部分设计则为语言简单而专业复杂，侧重于专业词汇的积累和语言技能的灵活应用，即先依托专业学科学习英语，后应用英语知识来拓展专业知识。此外，本教材充分利用移动互联的即时信息共享能力和多媒体的视听效果，有效借助理工科学习者的理性思维来提升学习效果，同时保留了听力、语感的基础性语言能力的训练，在一定程度上改变了教师在教学中的中心角色，降低了对教师专业背景和语言教学能力的门槛和要求，化解了在师资选择上的难题。在任务驱动的学习过程中，学习者能变被动积累知识为主动选择和吸收知识，因需而学，学以致用。

本教材和教学设计实现了重点突出、特色鲜明的特点，主要体现为以下几个方面。

第一，在不破坏知识系统性的基础上突出专业英语的特殊性。专业英语必须满足语言表达的客观性和准确性，因此对词汇的准确理解、使用和限定就成了教学的重点，特别是对词汇的辨析和对长句的阅读理解。

第二，突出能力。突出训练过程而不止步于学习素材，融入训练手册和学习指南，变知识的盲目被动积累为知识的灵活选择和主动吸收。

第三，突出学习者的中心地位。充分调动和发挥学习者的主观能动性，学习者可以自主地、灵活地、有重点和偏好地制订学习目标，选择学习内容、学习形式、学习时段和学习强度。教师成为学习的组织、管理和考核者，引导、帮助、监督和激励学习者完成学习任务。

第四，突出形式的多样化。包括内容承载形式的多样化（多媒体）和学习渠道的立体化（超媒体），兼顾趣味性和严谨性，以适当的形式承载适当的内容。

第五，突出英语学习，兼顾专业拓展，力求专业知识储备和语言能力同步提高。

第六，突出听、写的重要性。人类的信息交流是从语音开始的，对语音的记忆和识别是人类天生的能力。本教材设计了听、写练习，高度重视和充分利用上述能力。

第七，突出理性思考。理工科学生在学习英语时可以充分发挥理性思考的能力，尤其是对专业英语的学习。

第八，重视语言自身的规律。语言学习是一个从不断纠错到深刻和生动的过程，重要的知识和技能点需要被反复训练和强化。

本教材的编著是一次消除传统专业英语教学诟病的尝试，在教材设计与媒体制作中，英语教师和专业课教师深度合作，以计算机网络专业为例来落实先进的教学理念，应用现代化的教育技术和手段，以一种新形态、立体化的方式来实践"计算机网络专业英语"课程的教学，力求提高教材的作用，减轻专业课教师或英语教师的备课压力，降低教师在专业或语言方面的任教门槛。本教材以学生为中心面向自主学习，大学专科、职业院校和本科院校都可以通过本教材找到各自的学习切入点和最终目标，社会人士可以通过本教材进行自学。在认真完成教学设计中的学习任务后，英语语言知识、学习能力可望有大幅度提升，在不知不觉中提高专业水平。

　　本教材对计算机网络进行了系统的介绍，保持了专业知识的完整性，由浅入深地介绍了计算机网络的原理、应用、工程、管理，对当今的前沿，如物联网、云计算和大数据等也有所涉及，对英语语义及语法难点、重点专业知识、行业背景都进行了详尽的中英文注解。

　　建议在高等院校计算机网络及相关专业的"专业英语"课程或"计算机网络专业英语"双语课程中使用本教材，一般将其设计为 64 课时以上的综合性专业基础课程。

编　著　者

2024 年 5 月

目　　录

Professional English Learning Foundation
专业英语学习基础

1. 课程介绍

专业英语学习有哪三个层次的目标？

_____L0-1.MP4

2. 如何使用教材

本课程的学习主要需要完成哪些任务？

_____L0-2.MP4
_____

3. 理解和掌握专业英语的特点

专业英语的特点是什么？

_____L0-3.MP4

4. 对词汇学习的建议

语义和语意有什么区别？

_____L0-4.MP4

5. 词汇学习示例 1

举例说明如何通过词素分析来达到快速掌握词汇的目的。

_____L0-5.MP4
_____

6. 扩充专业词汇

扩充专业词汇的方法有哪些？

_____L0-6.MP4
_____

7. 词汇学习示例 2

专业英语学习的两个任务是什么？描述包含和从属关系的词汇有哪些？

_____L0-7.MP4
_____

8. 词汇学习示例 3

科技英语的生动性和准确性,哪个更重要?

L0-8.MP4

9. 长句分析示例 1

如何把握一个词的准确含义?

L0-9.MP4

10. 长句分析示例 2

专业英语常使用被动语态和虚拟语气吗?为什么?

L0-10.MP4

11. 长句分析示例 3

常用什么符号来简洁地表示补充说明?

L0-11.MP4

12. 长句分析示例 4

在遇到长句时,是否需要按照中文习惯调整语序?

L0-12.MP4

Unit One   Network and Its Application
网络及其应用

Lesson 1   Overview of Computer Network

Module 1   Text Study 课文学习

Basic Training 基本训练

Text 1   根据语音和视频完成以下任务。

Task 1-0 听录音，记录关键词，理解课文大意。

L1-1.MP3

Task 1-1 表示"网络"含义的词汇有哪些？语义有什么区别？
_____

_____ L1-1.MP4

Task 1-2 从专业的角度来看，network 和 computer network 之间有什么关系？
_____

Task 1-3 举例说明技术术语（专业词汇）和借用词汇的区别。
_____

Task 1-4 举例说明什么是复合词（合成词）。
_____

Task 1-5 分别以 cabling 和 linking 为例，说明什么是名词化，在什么时候需要名词化。
_____

Task 1-6 以下每组词汇的语义有什么区别？
① link、connect、attach_____
② link、linkage_____
③ connection、connectivity_____
④ device、equipment 和 facility_____

Task 1-7 网络中 link 交汇的地方叫什么（中、英文）？ _____

Task 1-8 signal、data、information、code、knowledge、intelligence 和 wisdom 彼此之间有什么关联？
_____

_____

Task 1-9 在课文中提到了哪些网络资源（用中、英文回答）？
_____

Task 1-10 从课文中找出通过前缀派生的词，并列举你知道的其他词。
_____

Task 1-11 以书写或口述录音的方式尝试复述课文，并检查和修正，建议进行多次，直至完全正确。
_____

_____

Task 1-12 effective 和 effect、efficient 和 efficiency 的语义有什么不同？
_____

Task 1-13 分析 China Telecom 和 telecommunication 的构词。
_____

Task 1-14 access 和 ACCESS 在语义上有什么不同？
_____

课文

The Definition of Computer Network

Computer network is a linking of the network devices and independent computers with each other to provide data communications and share network resources such as printers, scanners, modems, CD/DVD and internet access. Today, no business can work effectively without the data communications within the organization.

Text 2 根据语音和视频完成以下任务。

Task 2-0 听录音，记录关键词，理解课文大意。

_____

_____

L1-2.MP3

Task 2-1 在 Text 2 的文本上标注定语从句和宾语从句。

Task 2-2 从词源上分析 design 和 devise 语义的不同。

_____

L1-2.MP4

Task 2-3 从哲学层面分别解释 physical 和 physic、logical 和 logic 的语义。

_____

Task 2-4 比较 common、usual、normal、regular、ordinary、general 和 generic 的语义区别。

_____

_____

Task 2-5 Text 2 中表示"出现的频度"短语还有哪两个？

Task 2-6 比较 path、cable、wire、line、cord、channel、backbone 和 trunk 在词义上的区别。

_____

_____

Task 2-7 在对概念进行解释、比较时，常用的表达方式有哪些？

_____

Task 2-8 简述下列有关图、表的词汇的语义要点。

picture、figure、drawing、topology、graph、image、chart、map、diagram、photograph。

_____

_____

_____

_____

课文

Physical and Logical Layout of Computer Network

A network can be designed by different layouts known as topologies. Physical topology is the physical layout of the network, which defines how the cables are arranged and how the computers are connected. Logical topology refers to the nature of the paths the signals follow from node to node. In many instances, the logical topology is the same as the physical topology, but this is not always the case. The common topologies include bus, star, tree, mesh, hybrid and ring, as shown in Figure 1-1.

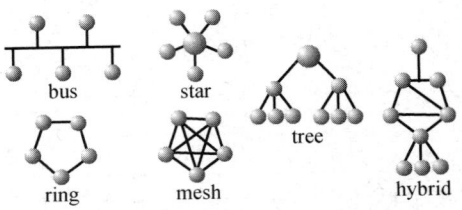

Figure 1-1   Network Topologies

Text 3   根据语音和视频完成以下任务。

Task 3-0 听录音，记录关键词，理解课文大意。

L1-3.MP3

_____

_____

Task 3-1 endpoint 或 end-point 在构词法中属于_____，由_____和 _____构成，还可以写成名词连用的形式_____。

L1-3.MP4

Task 3-2 理解、比较与翻译。

（1）endpoint、terminator、terminal、termination 和 terminate 的词义和词性分别是什么？

_____

（2）航站楼_____，登机口_____，gateway_____，《终结者》（电影名）_____。

（3）simultaneous（同时的）和 synchronous/synchronized（同步的）的语义有什么区别？

_____

（4）transmission（*n.*）/transmit（*v.*）和 transfer（*v.&n.*）都有"传输"的意思，它们在语义上的细微差别是_____。

（5）机场的"转机"和财务的"转账"分别用的是哪个词？_____

（6）比较下列词汇。

① receive & accept_____

② refer to & refer to … as_____

③ normally & commonly_____

Task 3-3 用下列句型造句。

① … cause …

_____

② …, which …

_____

③ … prevent … from …

_____

Task 3-4 前缀 "de-" 在单词 "degrade" 中的含义是_____。

Task 3-5 virtual 的反义词是_____

_____。

课文

Bus Topology

In bus topology, all of the nodes of the network are connected to a common transmission medium, which has exactly two endpoints (this is the "bus", which is also commonly referred to as the backbone or trunk). All data that is transmitted between nodes in the network is transmitted through this common transmission medium and is able to be received by all nodes in the network virtually simultaneously. The two endpoints of the common transmission medium are normally terminated with a device called a terminator that exhibits the characteristic impedance of the transmission medium and which dissipates or absorbs the energy that remains in the signal to prevent the signal from being reflected and propagated back onto the transmission medium in the opposite direction, which would otherwise cause interference and degradation of the signals on the transmission medium.

Text 4　根据语音和视频完成以下任务。

Task 4-0 听录音，记录关键词，理解课文大意。

_____　L1-4.MP3

_____

_____

Task 4-1 P2P 有歧义，例如，可能是_____或_____的缩写。　L1-4.MP4

Task 4-2 表示 "式样、方式、方法、模式" 的词汇有哪些？

_____

Task 4-3 翻译：基于以太网的_____。

Task 4-4 写出三个以 "re-"（表示 "再次"）为前缀的派生词_____。

Task 4-5 表示 "区域" 常用的词有 area、domain 和 region，它们在语义上的区别分别是什么？

_____

Task 4-6 用下列词汇造句，表示 "how to do something"。

① way_____

② mode_____

③ model_____

④ manner_____

⑤ pattern_____

⑥ fashion_____

⑦ form_____

⑧ type_____

Task 4-7 converge 的同（近）义词是_____。

Task 4-8 写出五个用"-based"合成的词。_____

Task 4-9 从计算机网络领域的文献中找出分别包含 area、domain 和 region 的例句。

① area_____

② domain_____

③ region_____

课文

Star Topology

In star topology, each node of the network is connected to a central node with a point-to-point link in a "hub" and "spoke" fashion, the central node being the "hub" and the nodes that are attached to the central node being the "spokes" (e.g. a collection of point-to-point links from peripheral nodes converging at a central node) — data transmitted between nodes in the network is transmitted to this central node, which usually retransmits the data to some or all other nodes in the network, although the central node may also be a simple common connection point without any active device to repeat the signals. Star topology is the most common topology in Ethernet-based Local Area Network.

Text 5　*根据语音和视频完成以下任务。*

Task 5-0 听录音，记录关键词，理解课文大意。

_____

_____ L1-5.MP3

Task 5-1 常用来描述数据本身"流动"的词是_____和_____。

Task 5-2 circular、circularity、cycled 和 cycle 的词性和词义分别是什么？

_____ L1-5.MP4

类似的词还有 ring 和 loop，网络中的"回路"该用哪个词？_____

Task 5-3 词汇翻译。

单向_____或_____；

双向_____；

多向_____；

反向_____。

Task 5-4 常用于叙述顺序的词是 the first、the next 和_____等。

Task 5-5 列出 generally 的一些近义词。

_____

课文

Ring Topology

In ring topology, each node of the network is connected to two other nodes in the network and with the first and the last nodes being connected to each other, a ring-all data that is transmitted between nodes in the network travels from one node to the next node in a circular manner, and the data generally flows in a single direction only.

Text 6　根据语音和视频完成以下任务。

Task 6-0 听录音，记录关键词，理解课文大意。

_____

_____　L1-6.MP3

_____

_____

Task 6-1 和 physical 对应的是 logical，和 practical（实践意义上的）对应的是 theoretical（理论意义上的），physical 和 practical 的语义分别是什么？

_____　L1-6.MP4

Task 6-2 specific、special 和 fixed 的语义区别是什么？

_____

Task 6-3 从 Lesson 1 中找出所有以 "inter-" 为前缀的词。_____

Task 6-4 主干网络用 backbone 和 trunk，分支网络用_____。

Task 6-5 请写出表示 "包含" 和 "组成" 的词汇和句型。

_____

_____

Task 6-6 分析 symmetrical 的构词。

_____

Task 6-7 翻译。

代价_____；性能_____；昂贵的_____或_____；性价比_____。

Task 6-8 i.e.如何发音和翻译？

_____

课文

Mesh，Tree (or Hierarchical) and Hybrid Topologies

In mesh topology, each node of the network is connected to each of the other nodes in the network with a point-to-point link—this makes it possible for data to be simultaneously transmitted from any single node to all of the other nodes. The physical fully connected mesh topology is generally too costly and complex for practical networks, although the topology is used when there are only a small number of nodes to be interconnected.

In tree topology, a central "root" node (the top level of the hierarchy) is connected to one or more other nodes that are one level lower in the hierarchy (i.e., the second level) with a point-to-point link between each of the second level nodes and the top level central "root" node, while each of the second level nodes that are connected to the top level central "root" node will also have one or more other nodes that are one level lower in the hierarchy (i.e., the third level) connected to it, also with a point-to-point link, the top level central "root" node being the only node that has no other node above it in the hierarchy—the hierarchy of the tree is symmetrical, each node in the network having a specific fixed number "f", being referred to as the "branching factor" of the hierarchical tree.

Hybrid topology is a type of network topology that is composed of one or more interconnections of two or more networks that are based upon different physical topologies mentioned above.

Text 7　*根据语音和视频完成以下任务。*

Task 7-0　听录音，记录关键词汇，理解课文大意。

_____

_____　L1-7.MP3

_____

Task 7-1　在 Text 7 中补充说明一个概念的方法是_____。

Task 7-2　缩写词（abbreviation, *abbr.*）用_____来标注，还原 UTP 是　L1-7.MP4

_____，缩写词一般如何发音？ _____

Task 7-2　从 Text 6 和 Text 7 中找出 level 和 layer，虽然都可翻译成"层"，但它们在语义上的差别是_____。

Task 7-3　填空。

The transmission media _____ electrical cable (UTP/STP _____, telephone _____, coaxial _____, and power _____), _____ optics and radio _____.

Task 7-4　用构词法分析 repeater（中继器）和 transceiver（收发器）在专业意义上的不同。

_____

Task 7-5　用 apart from 和 such as 造句。

① apart from_____

② such as_____

课文

The Physical Comprising

The transmission media used to link devices to form a computer network include electrical cable (UTP/STP cables, telephone wires, coaxial cables, and power lines), fiber optics and radio waves. In the OSI model, these are defined at layers 1 and 2—the Physical Layer and the Data Link Layer. Apart from any physical transmission medium, networks comprise additional basic devices, such as Network Interface Controllers (NICs), repeaters, hubs, transceivers, bridges, switches, routers, modems, and firewalls.

Text 8　*根据语音和视频完成以下任务。*

Task 8-0　听录音，记录关键词，理解课文大意。

_____

_____　L1-8.MP3

_____

Task 8-1　suite 的词义是_____，stack 的词义是_____，但 TCP/IP suite 和　L1-8.MP4

TCP/IP stack 在翻译时可以不区分。

Task 8-2　翻译。

Intranet_____　Extranet_____　internetwork_____　Internet_____

Task 8-3 "One of these specialized standards…" 中的 specialized 如何翻译？ _____

Task 8-4 填空。

① TCP/IP_____the protocol suite that_____multiple communication protocols, _____TCP, IP, SMTP, FTP, DHCP, LDAP, PPP, Telnet_____.

② No computers_____the Internet can communicate without a_____IP address.

③ There are three main types of Internet works_____Intranet, Extranet and Internet.

④ The communications on the Internet_____the IP address, which is a_____of the TCP/IP stacks.

Task 8-5 common 在课文中出现过两次，分别翻译成什么？

_____

_____

Task 8-6 用英语解释什么是 communication protocol。

_____

_____

课文

Protocols

One of these specialized standards and agree-up on ways is known as protocols. The TCP/IP is the protocol suite that contains multiple communication protocols, such as TCP, IP, SMTP, FTP, DHCP, LDAP, PPP, Telnet and many others. TCP/IP works together and is the most common communication protocol for LAN, WAN and Internet. Communications on the Internet are based on the IP addresses, which are part of the TCP/IP stacks. No computers on the Internet or in the LAN/MAN/WAN can communicate without a unique IP address. An internetwork is a type of computer network that connects two or more different networks via a common routing technology, using routers. There are three main types of internetworks, i.e., Intranet, Extranet and Internet.

Advanced Training 进阶训练

T1.PDF

Task 1 根据段落内容和中文提示填空。

NOS

In addition, a network can be set up by computers with Network Operating System (NOS) _____（例如）Windows/Linux installed. Every computer requires a _____（唯一的）LAN card, which should be properly installed and configured, and Ethernet cable with the RJ-45 connectors at both ends. To communicate with the other computer, each computer should support the same protocol and the TCP/IP is the most _____（通常地）used protocol in a computer network. Each network is either a peer-to-peer or client/server network. In a _____（客户/服务）network _____（模式）, a centralized domain controller is used to control all the computer networks. It provides the services to the clients like logon, authentication, printer _____（访问），_____（中心化的）data storage, user _____（管理），resources management, DHCP, DNS, FTP and internet access.

Task 2 用下列词汇或词组填空。

global、private、public、scope、consists、an extensive range of、virtually、origins、back to、robust、fault-tolerant、exponential、initially、segment、underpinning、principal、non-profit、expertise。

Internet

The Internet is a _____ （全球化）system of interconnected computer networks that use the Internet protocol suite (TCP/IP) to link billions of devices worldwide. It is a network that _____ （包含）of millions of _____ （私有的）, _____ （公有的）, academic, business, and government networks of local to global _____ （范围）, linked by a broad array of electronic, wireless, and optical networking technologies.

The Internet carries _____ （广泛的）information resources and services, such as the interlinked hypertext documents and applications of the World Wide Web (WWW), electronic mail, telephony, and peer-to-peer networks for file sharing.

Although the Internet protocol suite has been widely used by academia and the military industrial complex since the early 1980s, events of the late 1980s and 1990s such as more powerful and affordable computers, the advent of fiber optics, the popularization of HTTP and the Web browser, and a push towards opening the technology to commerce eventually incorporated its services and technologies into _____ （事实上）every aspect of contemporary life.

The _____ （起源）of the Internet date _____ （追溯到）research and development commissioned by the United States government, the Government of the UK and France in the 1960s to build _____ （健全的）, _____ （容错的）communication via computer networks. This work led to the primary precursor networks, the ARPANET. The interconnection of regional academic networks in the 1980s marks the beginning of the transition to the modern Internet. From the late 1980s onward, the network experienced sustained _____ （指数的）growth as generations of institutional, personal, and mobile computers were connected to it.

Most traditional communication media, including telephony and television, are being reshaped or redefined by the Internet, giving birth to new services such as Internet telephony and Internet television. Newspaper, book, and other print publishing are adapting to website technology, or are reshaped into blogs and Web feeds. The entertainment industry was _____ （最初地）the fastest growing _____ （部分）on the Internet. The Internet has enabled and accelerated new forms of personal interactions through instant messaging, Internet forums, and social networking. Online shopping has grown exponentially both for major retailers and small artisans and traders. Business-to-business and financial services on the Internet affect supply chains across entire industries.

The Internet has no centralized governance in either technological implementation or policies for access and usage; each constituent network sets its own policies. Only the overreaching definitions of the two _____ （主要的）name spaces in the Internet, the Internet Protocol address space and the Domain Name System (DNS), are directed by a maintainer organization, the Internet Corporation for Assigned Names and Numbers (ICANN). The technical _____ （基础）and standardization of the core protocols is an activity of the Internet Engineering Task Force (IETF), a _____ （非营利）organization of loosely affiliated international participants that anyone may associate with by contributing technical _____ （专门知识）.

Vocabulary Practice 词汇练习

词汇及短语听写，并纠正发音。

C1.MP3

_____

_____

_____

_____

Words & Expressions

layout [ˈleɪaʊt]	n. 布局，安排，设计；布置图，规划图
node [nəʊd]	n. 节点；（计算机网络的）节点
transmit [trænsˈmɪt]	vi. 传输；发射；传送，传递
simultaneously [ˌsɪm(ə)lˈteɪnɪəslɪ]	adv. 同时地
impedance [ɪmˈpiːdns]	n. 电阻抗；阻抗，全电阻
dissipate [ˈdɪsɪpeɪt]	vi. 驱散；（使）消散；浪费，挥霍
propagate [ˈprɒpəgeɪt]	vt. 传播；宣传；繁衍；遗传
degradation [ˌdegrəˈdeɪʃn]	n. 毁坏，恶化，堕落，潦倒
hub [hʌb]	n. 轮轴；中心；枢纽
spoke [spəʊk]	n. 轮辐，（车轮的）辐条
peripheral [pəˈrɪfərəl]	adj. 次要的；附带的；外围的
converge [kənˈvɜːdʒ]	vi. 聚集；（线条、运动的物体等）汇聚于一点
interconnect [ˌɪntəkəˈnekt]	v. 使互相连接，连接
hierarchical [ˌhaɪəˈrɑːkɪkl]	adj. 层次结构的；按等级划分的
symmetrical [sɪˈmetrɪkl]	adj. 对称的，匀称的
coaxial [kəʊˈæksɪəl]	adj. 同轴的，共轴的
comprise [kəmˈpraɪz]	vt. 包含；由……组成
authentication [ɔːˌθentɪˈkeɪʃn]	n. 认证；身份验证；证明
array [əˈreɪ]	n. 数组；队列，阵列；一大批
optical [ˈɒptɪkl]	adj. 视力的；视觉的；光学的
advent [ˈædvent]	n. 出现；到来
incorporate [ɪnˈkɔːpəreɪt]	vt. 吸收，（使）混合；合并；包含
contemporary [kənˈtemp(ə)rərɪ]	adj. 当代的，现代的；同时代的
robust [rəʊˈbʌst]	adj. 强健的；健康的
precursor [priːˈkɜːsə(r)]	n. 前驱，先锋；初期形式
exponential [ˌekspəˈnenʃl]	adj. 越来越快的，迅猛增长的；指数的，幂数的
retailer [ˈriːteɪlə(r)]	n. 零售商，零售店
artisan [ˌɑːtɪˈzæn]	n. 技工，工匠

governance [ˈɡʌvənəns]　　　　　　*n.* 管理；统治；支配
constituent [kənˈstɪtjuənt]　　　　*adj.* 构成的，组成的
affiliated [əˈfɪlɪeɪtɪd]　　　　　　*adj.* 隶属的；附属的
cable [ˈkeɪbl]　　　　　　　　　　*n.* 缆绳，绳索；电缆

Module 2　Study Aid 辅助帮学

Terms & Abbreviations 术语和缩写

bus　　总线（在信息科学中），意为公用、共享的信息传输通道。

topology　　　拓扑、拓扑学。拓扑学是几何学的一个分支。拓扑是网络拓扑图和网络拓扑结构的简称，指网络的几何形态，可以是物理（地理）或逻辑意义上的。

transmission medium　　　传输媒体或者传输介质，是数据传输系统中物理信号的载体和通道，分为导向传输媒体（有线）和非导向传输媒体（无线）。

Ethernet　　以太网，是指以多路访问共享信道为基础的计算机网络。

UTP　　*abbr.* 非屏蔽双绞线（Unshielded Twisted Pair），是目前最常用的短距离（100 米左右）数字通信线缆之一，一般用于户内。

STP　　*abbr.* 屏蔽双绞线（Shielded Twisted Pair），也是目前最常用的短距离（100 米左右）数字通信线缆之一，一般用于户外。

OSI　　*abbr.* 开放式系统互联（Open System Interconnection），是国际标准化组织（ISO）制定的网络互联结构的参考模型，该模型定义了计算机互联的标准，是设计和描述计算机网络通信的基本框架。

NICs　　*abbr.* 网络适配器（Network Interface Controllers），是计算机连接局域网的电路模块。

transceiver　　收发器，指用于转发数据包的电子设备，特别指光纤收发器。

TCP/IP　　*abbr.* 传输控制协议/互联网协议（Transmission Control Protocol / Internet Protocol），也称网络通信协议，是 Internet 最基本的协议、Internet 国际互联网络的基础，由网络层的 IP 和传输层的 TCP 两个协议组成。

protocol suite　　协议族。协议族和协议栈（protocol stack）都是指一组相关的协议，例如，TCP/IP 协议族包含 TCP、IP、SMTP、FTP、DHCP、LDAP、PPP 和 Telnet 等协议。

SMTP　　*abbr.* 简单邮件传输协议（Simple Mail Transfer Protocol），是一组用于从源地址到目的地址传输邮件的规则，用来控制邮件的中转方式。

FTP　　*abbr.* 文件传输协议（File Transfer Protocol），用于控制文件在 Internet 上的双向传输。

DHCP　　*abbr.* 动态主机配置协议（Dynamic Host Configuration Protocol），在局域网中用于给计算机自动分配 IP 地址，并作为管理员的中央管理手段。

LDAP　　*abbr.* 轻量级目录访问协议（Lightweight Directory Access Protocol），是关于从目录中创建、访问和移除对象和数据的一种 IETF 协议。

PPP　　*abbr.* 点对点协议（Point-Point Protocol），是一个被广泛使用的广域网协议，跨过同步和异步电路实现路由器到路由器（Router-to-Router）和主机到网络（Host-to-Network）

的点对点连接。PPP 也可以代指端到端（Peer-Peer Protocol）协议。

Telnet   远程登录协议，是 TCP/IP 协议族中的一员，是 Internet 远程登录服务的标准协议和主要方式，为用户提供了在本地计算机上完成在远程主机控制台（键盘）上工作的能力。

LAN   *abbr.* 局域网（Local Area Network），又称内网，是一个在局部的地理和行政管理范围内（如一所学校、一家工厂，一般是方圆几千米），将各种计算机、外部设备和数据库等互相连接起来组成的计算机通信网，以广播通信和共享信道为基础。

WAN   *abbr.* 广域网（Wide Area Network），也称远程网（Long Haul Network），通常跨接很大的物理范围和不同的行政管理区域，所覆盖的范围为几十千米到几千千米，它能连接多个城市、国家，或横跨几个洲，并能提供远距离通信，形成国际性的远程网络。

MAN   *abbr.* 城域网（Metropolitan Area Network），是在城市范围内建立的计算机通信网，属于宽带局域网。

Intranet   企业内部网，或称内部网、内联网、内网，是一个使用与互联网同样技术的计算机网络，它可以说是 Internet 技术在企业内部的应用。

Extranet   外联网，是一个使用 Internet 技术使企业与客户、其他企业相连以完成共同目标的合作网络。

Internet   互联网，又叫国际互联网，是目前正在运行的最基本和通用的全球性互联网络，以 TCP/IP 为基础。

NOS   *abbr.* 网络操作系统（Network Operating System），指专为提供网络服务而开发的操作系统版本，利用低层网络提供的数据传输功能为高层网络用户提供各种服务，具有多用户、多线程管理的能力，相对于桌面操作系统，强化了安全（Security）及并发（Concurrency）性能。

configure   配置（这里是动词，它的名词为 configuration，简写为 config），指对程序参数的设置，通常以纯文本格式保存，例如，config.sys、config.ini。

logon   登录，和 login 是同义词，反义词是 logout（退出、注销）。

DNS   *abbr.* 域名系统（Domain Name System），是互联网上作为域名和 IP 地址相互映射的一个分布式数据库系统，使用户更方便地通过域名访问互联网。域名通常具有被人直接理解的含义而方便记忆，但 IP 地址却不是，虽然计算机最终仍是用 IP 地址来实现路由的。

WWW   *abbr.* 万维网（World Wide Web），是环球信息网的缩写，简称 Web，它是互联网中最基本和使用最广泛的一个应用，该应用直接推动了互联网的迅猛发展。

HTTP   *abbr.* 超文本传输协议（Hypertext Transfer Protocol），是互联网中应用最为广泛的一种网络协议，所有的 WWW 文件都必须遵守这个协议。

ARPANET   *abbr.* 阿帕网（Advanced Research Project Agency NETwork），是互联网的起源，1969 年诞生在美国，最初是高等研究计划局开发的一个军用专属网络。

ICANN   *abbr.* 互联网名称与数字地址分配机构（Internet Corporation for Assigned Names and Numbers），该机构成立于 1998 年，是一个集合了全球网络界、商业、技术及各学术领域专家的非营利性国际组织，负责互联网唯一标识符系统及其安全、稳定运营的协调。

IETF   *abbr.* 互联网工程任务组（Internet Engineering Task Force），成立于 1985 年，是全球互联网最具权威性的技术标准化组织，主要任务是负责互联网相关技术标准规范的研发和制定。

Difficult Sentences Analysis and Translation 难句分析与翻译

1. The two endpoints of the common transmission medium are normally terminated with a device called a terminator that exhibits the characteristic impedance of the transmission medium and which dissipates or absorbs the energy that remains in the signal to prevent the signal from being reflected and propagated back onto the transmission medium in the opposite direction, which would otherwise cause interference and degradation of the signals on the transmission medium.

译文：公共传输介质的两个端点通常由一个装置来终止，该装置称作终端器。终端器匹配传输介质的特性阻抗，消耗或吸收信号中留存的能量以防止信号被反射，从相反方向传播回传输介质，否则，就会对传输介质上的信号造成干扰和衰减。

分析：句中过去分词短语"called a terminator"作定语，修饰、限定其前面的名词"device"；"that exhibits … the transmission medium"和"which dissipates or absorbs … in the opposite direction"都是定语从句，修饰、限定其前面的名词"terminator"；定语从句"that remains in the signal"修饰、限定其前面的名词"energy"；"which would … cause interference … on the transmission medium"是非限制性定语从句虚拟语气，对前半句做补充说明。

2. Text 4 全段

译文：在星型拓扑中，网络的每个节点都以"轮毂"和"辐条"的方式和中央节点进行点对点的连接，中央节点为"轮毂"，而连接到中央节点的各节点为"辐条"（例如，汇集在中央节点外围的点对点链路集合）——在网络中的节点之间传输的所有数据都被传输到中央节点。尽管中央节点也可能是一个简单的公共连接点，并没有任何用以中继信号的有源设备，但该中央节点通常是某种能够将数据重新发送到网络中一些或所有其他节点的设备。

分析：本段结构较为复杂，可以通过断句重整来理解。句中介词短语"with a point-to-point link in a 'hub' and 'spoke' fashion"作定语，修饰、限定其前面的整个句子的主体结构；"the central node being the 'hub'"和"the nodes that are attached to the central node being the 'spokes'"都是分词独立结构作状语；在破折号后面，过去分词短语"transmitted between nodes in the network"修饰、限定其前面的"data"；非限制性定语从句"which usually retransmits the data to some or all of the other nodes in the network"对前半句做补充说明；"although the central node may also be a simple common connection point…"是让步状语从句。

3. The physical fully connected mesh topology is generally too costly and complex for practical networks, although the topology is used when there are only a small number of nodes to be interconnected.

译文：虽然拓扑只在有少量节点互联时可用，但是这种物理全连接网状拓扑在实际的网络运用中结构很复杂，费用也很高。

分析："although the topology is used …"是让步状语从句；"when there are only a small number of nodes to be interconnected"是时间状语从句。

4. Hybrid Topology is a type of network topology that is composed of one or more

interconnections of two or more networks that are based upon different physical topologies mentioned above.

译文：混合拓扑是一种网络拓扑，它由一个或多个建立在上述不同物理拓扑的网络互联而组成。

分析：本句中 "that is composed of one or more interconnections of two or more networks" 是定语从句，修饰、限定 "network topology"，其后的 "that are based upon different physical topologies mentioned above" 也是定语从句，修饰、限定前面的 "two or more networks"；过去分词短语 "mentioned above" 作定语，修饰、限定 "different physical topologies"。

5. It is a network that consists of millions of private, public, academic, business, and government networks of local to global scope, linked by a broad array of electronic, wireless, and optical networking technologies.

译文：这是一个由数以百万计的从地方到全球范围内的私有、公共、学术、商业和政府网络组成的网络。该网络由大量电子、无线和光网络技术连接。

分析：本句中短语 "a broad array of" 的意思是 "广泛的、一系列的、大量的"；过去分词短语 "linked by a broad array of ... technologies" 作后置定语，说明该网络 "由大量电子、无线和光网络技术连接"，"that consists of millions of ... to global scope" 是定语从句，修饰、限定 "networks"。

6. Although the Internet protocol suite has been widely used by academia and the military industrial complex since the early 1980s, events of the late 1980s and 1990s such as more powerful and affordable computers, the advent of fiber optics, the popularization of HTTP and Web browser, and a push towards opening the technology to commerce eventually incorporated its services and technologies into virtually every aspect of contemporary life.

译文：虽然互联网协议族自 20 世纪 80 年代初就已经广泛应用于学术界和军工企业，但是 20 世纪 80 年代末和 90 年代发生的事件，如更强大而平价的计算机、光纤的问世，HTTP 和 Web 浏览器的普及，以及技术向商务开放的推动等，才最终将其服务和技术几乎融入了当代生活的各个方面。

分析：本句中 "Although ..." 引导让步状语从句；短语 "incorporate ... into ..." 意思是 "把……纳入……，使……并入……"。

7. Only the overreaching definitions of the two principal name spaces in the Internet, the Internet Protocol address space and the Domain Name System (DNS), are directed by a maintainer organization, the Internet Corporation for Assigned Names and Numbers (ICANN).

译文：只有互联网的两个主要命名空间（互联网协议地址空间和域名系统）的根定义是由一家维护机构——互联网名称与数字地址分配机构（ICANN）来指定的。

分析：句子以 "Only" 开头，强调了主句的内容。

8. The technical underpinning and standardization of the core protocols is an activity of the Internet Engineering Task Force (IETF), a non-profit organization of loosely affiliated international participants that anyone may associate with by contributing technical expertise.

译文：互联网工程任务组（IETF）的一项任务就是制定核心协议的技术基础和标准规范。该任务组是一个非营利性组织，接纳松散关联的国际会员，任何人均可以通过贡献专业技术成为会员。

分析：本句中短语"a non-profit organization …"是同位语，补充说明 the Internet Engineering Task Force (IETF)；介词短语 "of loosely affiliated international participants" 修饰、限制其前的名词短语； "that anyone may associate with by contributing technical expertise" 是定语从句，修饰、限定"international participants"。

Module 3　Consolidation Exercise 巩固练习

I. Translate the following words and expressions into English

K1.PDF

网络_____ 连接_____ 设备_____ 通信_____

资源_____ 打印机_____ 扫描仪_____ 调制解调器_____

存取_____ 光纤_____ 无线的_____ 集线器_____

使专业化_____ 协议_____ 套件_____ 多重的_____

安装_____ 配置_____ 以太网_____ 连接器_____

路由器_____ 交换机_____ 网桥_____ 收发器_____

服务器_____ 集中的_____ 域_____ 控制器_____

登录_____ 认证_____ 存储_____ 互联网_____

II. Choose the best answer to the questions according to the passage

1. What functions does computer networking have? _____

A. Providing data communications.

B. Sharing network resources.

C. Serving the businesses effectively.

D. All of the above.

2. A computer network can be interconnected with _____.

A. modem

B. scanner

C. UTP/STP cable

D. mouse

3. Which is the most common topology in Ethernet-based Local Area Network? _____

A. Hub topology.

B. Mesh topology.

C. Star topology.

D. Ring topology.

4. Which of the following does NOT belong to TCP/IP suite? _____

A. FTP.

B. DVD.

C. Telnet.

D. DHCP.

5. Which of the following statements is NOT TRUE? _____

A. No computer on the Internet can communicate without a unique IP address.

B. TCP/IP is the most common communication protocol for LAN, WAN and Internet.

C. Every computer requires a unique LAN card, which should be properly installed and configured before the computer can join the network.

D. There are two main types of computer network.

III. Translate the following sentences into Chinese

1. Today, no business can work effectively without data communications within the organization.

_____

_____

2. TCP/IP works together and it is the most common communication protocol for LAN, WAN and Internet.

_____

_____

3. Every computer requires a unique LAN card, which should be properly installed and configured.

_____

_____

4. In topology, there are specialized rules and standards, and based on these standards, the devices communicate with each other.

_____

_____

5. To communicate with the other computer, each computer should support the same protocol.

_____

_____

6. Internetwork is a type of computer network that connects two or more different networks.

_____

_____

7. The communication on the Internet is based on the IP address, which is a part of the TCP/IP stack.

_____

_____

8. TCP/IP is the protocol suite that contains multiple communication protocols.

_____

_____

Lesson 2　Network Classification

Module 1　Text Study 课文学习

Basic Training 基本训练

Text 1　根据语音和视频完成以下任务。

可通过 Fignre 1-2 查看各网络的详细信息。

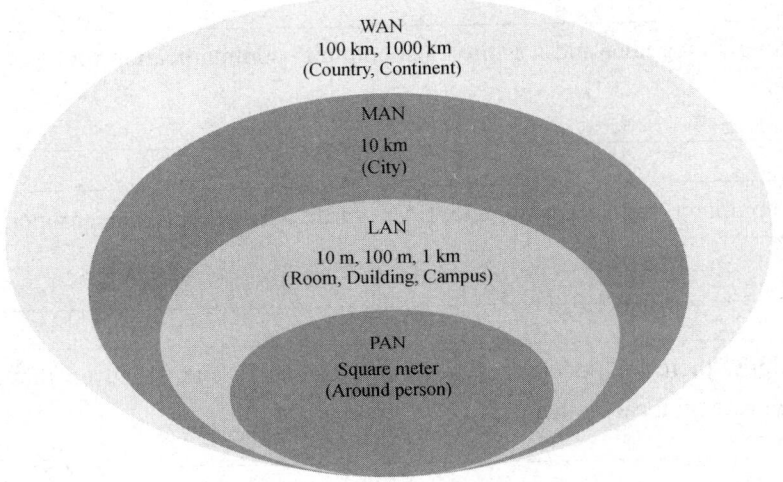

WAN
100 km, 1000 km
(Country, Continent)

MAN
10 km
(City)

LAN
10 m, 100 m, 1 km
(Room, Duilding, Campus)

PAN
Square meter
(Around person)

Figure 1-2　Network Size

Task 1-0 听录音，记录关键词，理解课文大意。

_____

_____

L2-1.MP3

Task 1-1 写出下列缩写词的中、英文含义。

① PAN_____

② LAN_____

L2-1.MP4

③ MAN_____

④ WAN_____

⑤ WLAN_____

⑥ VLAN_____

⑦ VPN_____

Task 1-2 表示标准、规则、定律、协议的词汇有哪些？它们在词义上分别有什么区别？

_____

_____

Task 1-3 写出词义最接近的英语词汇。

规模_____；大小_____；范围_____；区域_____；本地的_____；远程的_____。

Task 1-4 class、classify、categorize 和 category 在词性和语义上分别有什么区别？

_____

_____

课文

Overview of Network Classification

One important criterion for classifying networks is their scale (physical size). As shown in Figure 1-2, there are:

- Personal Area Network (PAN).
- Local Area Network (LAN).
- Metropolitan Area Network (MAN).
- Wide Area Network (WAN).

Text 2　根据语音和视频完成以下任务。

Task 2-0 听录音，记录关键词，理解课文大意。

_____

L2-2.MP3

_____

Task 2-1 翻译画线部分。

外围设备_____是主机的 CPU 不能直接读写其内存值的设备，须通过物理接口_____和相关通信协议_____，外围设备其实是一种　　L2-2.MP4

嵌入式_____计算机，连接____在主机上，以主从模式_____工作。

Task 2-2 etc.是拉丁语，表示_____，读音是_____，等于 and so on，"."不能省略。

Task 2-3 setting 在 Test 2 中的含义是与 configuration 还是 situation 相同？_____

Task 2-4 technology、technique 和 skill 都可翻译成_____，但它们的语义区别是：

_____

_____。

Task 2-5 RFID 是_____的缩写。

Task 2-6 翻译。

智能卡_____　智慧城市_____　人工智能_____

Personal_____　私有的_____　公有的_____

Task 2-7 翻译或填空。

Bluetooth can be used in other _____, too. It is often used to connect a headset to a mobile phone without_____（电线），and it can allow your digital music player to connect to your car _____（仅仅）being brought within _____. A completely different _____of PAN is _____（形成）when an embedded medical device talks to a user-operated remote control. PANs can also be _____（建造）with other _____（技术）.

课文

Personal Area Network (PAN)

Personal Area Networks let devices communicate over the range of a person. A common example is a wireless network that connects a computer with its peripherals. Almost every computer has an attached monitor, keyboard, mouse and printer. Without using a wireless network, this connection must be done with cables. So many new users have a hard time finding the right cables and plugging them into the right little hole. To help these users, some companies got together to design a short-range wireless network called Bluetooth to connect these components without wires. In the simplest form, Bluetooth networks use the master-slave paradigm. The system unit (the PC) is normally the master, talking to the mouse, keyboard, etc., as slaves. Bluetooth can be used in other settings, too. It is often used to connect a headset to a mobile phone without cords and it can allow your digital music player to connect to your car merely being brought within range. A completely different kind of PAN is formed when an embedded medical device such as a pacemaker, insulin pump, or hearing aid talks to a user-operated remote control. PANs can also be built with other technologies that communicate over short ranges, such as RFID on smartcards and library books.

Text 3   根据语音和视频完成以下任务。

Task 3-0 听录音，记录关键词，理解课文大意。

L2-3-1.MP3

_____

_____

_____

L2-3-2.MP3

_____

_____

_____

L2-3-3.MP3

_____

_____

Task 3-1 表示举例的词和词组有：such as、for example、_____和拉丁语的_____（请思考如何读）。

L2-3.MP4

Task 3-2 表示"限制"的词有_____、_____和_____。

Task 3-3 …that would not otherwise be possible…的中文意思是_____。

Task 3-4 课文中表示"运行"的词有_____、_____和_____。

Task 3-5 翻译。

……运行速度在 10 Mbps 到 100 Mbps 之间_____；

……运行速度高达 10 Gbps_____。

Task 3-6 表示"基础设施"的词有_____和_____。

Task 3-7 找出并翻译分别含有 accommodate、incorporate 和 correspond 的句子。

_____

_____

_____

Task 3-8 program 在课文中的含义是_____。

Task 3-9 数据链路层的数据交换单位是_____，在网络层则为_____。

Task 3-10 翻译。

The techniques for using a shared medium in this way were later applied to wired technology in the form of Ethernet.

Task 3-11 填空并指出所填写词汇的词义。

The first version of Ethernet _____ a media access method _____ Carrier Sense Multiple Access with _____ CSMA/CD.

课文

Local Area Network (LAN)

Local Area Networks are privately-owned networks covering a small geographic area, like a home, office, building or group of buildings (e.g., campus). They are widely used to connect computers in company offices and factories to share resources (e.g. printers) and exchange information. LANs are restricted in size, which means that the worst-case transmission time is bounded and known in advance. Knowing this bound makes it possible to use certain kinds of designs that would not otherwise be possible. It also simplifies network management. Traditional LANs run at speeds of 10 Mbps to 100 Mbps, have low delay (microseconds or nanoseconds), and make very few errors. Newer LANs operate at up to 10 Gbps. Ethernet is the predominant LAN technology in use today.

The foundation for Ethernet technology was first established in 1970 with a program called ALOHAnet. ALOHAnet was a digital radio network designed to transmit information over a shared radio frequency between the Hawaiian Islands. ALOHAnet required all stations to follow a protocol in which an unacknowledged transmission required re-transmitting after a short period of waiting. The techniques for using a shared medium in this way were later applied to wired technology in the form of Ethernet. Ethernet was designed to accommodate multiple computers that were interconnected on a shared bus topology. The first version of Ethernet incorporated a media access method known as Carrier Sense Multiple Access with Collision Detection (CSMA/CD). CSMA/CD managed the problems that arose when multiple devices attempted to communicate over a shared physical medium.

This topology became more problematic as LANs grew larger and LAN services made increasing demands on the infrastructure. The original thick coaxial and thin coaxial physical media were replaced by early categories of UTP cables. A significant development that enhanced LAN performance was the introduction of switches to replace hubs in Ethernet-based networks. This development closely corresponded with the development of 100BASE-TX Ethernet. Switches can control the flow of data by isolating each port and sending a frame only to its proper destination (if the destination is known), rather than sending every frame to every device.

Text 4   根据语音和视频完成以下任务。

Task 4-0 听录音，记录关键词，理解课文大意。

L2-4-1.MP3

_____

_____

_____

L2-4-2.MP3

_____

_____

Task 4-1 写出与中文最接近的英文单词，理解它们的不同含义。

最初的_____；原来的_____；初级的_____；原始的_____。

L2-4.MP4

Task 4-2 extend、expand 和 exceed 有共同的前缀_____，表示"向外"，它们的词义分别是：_____。

Task 4-3 complexity 和 connectivity 有共同的后缀_____，表示_____。

Task 4-4 表示"造"的最通用、最常用的一个词是_____，此外还有_____（建造、创立）、_____（形成、构成）、_____（建立、确定）和_____（创造）。

Task 4-5 _____和_____都表示"以……为基础"，其中_____是及物动词，_____是不及物动词。"根据、基于"表示抽象的基础，on the_____of。

Task 4-6 subnet、subsystem 都有前缀_____，表示_____。

Task 4-7 "It can now be applied across a city in what is known as a Metropolitan Area Network." 中，"It can be applied in a Metropolitan Area Network" 是主体结构，"across a city" 是状语，"what is known as" 是宾语从句。

完整的翻译为：_____。

Task 4-8 填空/翻译/替换。

① The Internet is the widest WAN built（或_____）yet.

② Wide Area Networks can be constructed depending on（或_____，或 on the basis of）what kind of linking or Internet connection is used.

③ Wide Area Network technology has three major constitute parts（或 elements 或_____）.

④ All computers with an_____（相同的）initial part in the IP address are on a_____（共同的）suburb.

Task 4-9 列举 5 个以 -ity 为后缀，表示"……性"的词汇_____。

Task 4-10 比较和体会 assign、allocate、distribute 和 delivery 的词义。

课文

Metropolitan Area Network (MAN) and Wide Area Network (WAN)

The increased cabling distances enabled by the use of fiber-optic cable in Ethernet-based networks has resulted in a blurring of the distinction between LANs and WANs. Ethernet was initially limited to LAN cable systems within a single building, and then extended to between buildings. It can now be applied across a city in what is known as a Metropolitan Area Network.

The Wide Area Network technology is a method of creating an intelligent data sharing network of computers, whose complexity and scale is so huge that it may extend over a city, a region, or a country. It may even be an international network. The Internet is the widest WAN constructed yet. Nowadays, private organizations like big software companies have their own WANs constructed. It makes their internal communication simpler and more secure. Internet service providers provide connectivity and infrastructure to private LANs and integrate them into the Internet. Wide Area Networks can be constructed in a variety of ways depending on what kind of linking or Internet connection is used. WANs could also be classified on the basis of what kind of Internet protocol or data sharing technique they use.

Wide Area Network technology has three major components. The main component is the data line that connects the network to the Internet superhighway. This may vary in the data transfer speed it offers. Then come to the routers, that is, the routing mechanism and the software component or protocol mechanism. Routers are the most important components in this technology. They are also called "Level 3 switches". Routers coupled with the data sharing protocols, form a huge data delivery system.

All data that is transmitted on the Internet is always labeled with the destination address which is a numerical name assigned to every computer connected on the network. They are called "Internet Protocol (IP) addresses". Routers decide bandwidth allocation for every type of data that is transmitted. Routers are placed at nodes of a network connecting many subnets together. A subnet is a network of computers with an identical initial part in the IP address. You could say that all those computers in a subnet are on a common suburb in the network distribution.

Advanced Training 进阶训练

Task 根据提示填空。   T2.PDF

Wireless Local Area Network (WLAN)

A _____（无线）Local Area Network (WLAN) is a wireless computer network that _____（连接）two or more devices using a wireless _____（分配 destri…）method within a _____（有限的）area such as a home, school, computer laboratory, or office building. This gives users the ability to move around within a local _____（覆盖 c…ge）area and still be connected to the network, and can provide a _____（连接…tion）to the wider Internet. Most modern WLANs are based on IEEE 802.11 standards, marketed under the Wi-Fi brand name.

WLANs have become popular in the home due to ease of _____（安装 in…tion）and use, and in commercial complexes offering wireless access to their customers, often for free. New York City, for instance, has begun a pilot program to provide city workers in all five boroughs of the city with wireless Internet access.

The IEEE 802.11 has two basic modes of operation: Infrastructure and ad hoc mode. In ad hoc mode, mobile units transmit directly _____（对等 peer-to-…）. In infrastructure mode, mobile units communicate through an access point that serves as a bridge to other networks (such as Internet or LAN).

Since wireless communication uses a more open medium for communication _____（相

比 in comp... on）to wired LANs, the 802.11 designers also included encryption mechanisms: Wired Equivalent Privacy (WEP, now insecure), Wi-Fi Protected Access (WPA, WPA2), to secure wireless computer networks. Many access points will also offer Wi-Fi Protected Setup, a quick (but now ＿＿＿＿＿（不安全的）) method of joining a new device to an ＿＿＿＿＿（加密的） network.

Vocabulary Practice 词汇练习

词汇及短语听写，并纠正发音。

_____ C2.MP3

_____

_____

_____

Words & Expressions

criterion [kraɪˈtɪərɪən]	n.（批评、判断等的）标准，准则；规范
component [kəmˈpəʊnənt]	n. 组件；成分；[数]要素
	adj. 成分的；组成的；合成的；构成的
frequency [ˈfriːkwənsɪ]	n. 频率，次数；频繁性，频率分布
accommodate [əˈkɒmədeɪt]	vt. 调解，调停；容纳，使适应；向……提供住处
multiple [ˈmʌltɪpl]	adj. 多个的，多重的；复杂的；多功能的
infrastructure [ˈɪnfrəstrʌktʃə(r)]	n. 基础设施；基础建设
installation [ˌɪnstəˈleɪʃn]	n. 安装；装置
complex [ˈkɒmpleks]	n. 综合体
	adj. 复杂的
borough [ˈbʌrə]	n. 作为行政区划的镇、区（主要适用于美国）
encryption [ɪnˈkrɪpʃn]	n. 加密；编密码
blurring [blɜːrɪŋ]	n.（图像的）混乱；模糊
complexity [kəmˈpleksətɪ]	n. 复杂性，复杂的状态；复杂的事物；复合物
connectivity [ˌkɒnekˈtɪvɪtɪ]	n. 连通性；连接
integrate [ˈɪntɪgreɪt]	vt.（使）成为一体；（使）整合；（使）完整
bandwidth [ˈbændwɪdθ]	n. 带宽；频宽
allocation [ˌæləˈkeɪʃn]	n. 分配额（或量）；配给，分配；划拨的款项
pacemaker [ˈpeɪsmeɪkə(r)]	n. 起搏器
insulin [ˈɪnsjəlɪn]	n. 胰岛素

Module 2　Study Aid 辅助帮学

Terms & Abbreviations 术语和缩写

PAN　　abbr. 个人局域网（Personal Area Network），指个人范围（随身携带或数米之内）的智能设备（如计算机、电话、PDA、数码相机等）组成的通信网络。

Bluetooth　　蓝牙，一种无线技术，可实现固定设备、移动设备和楼宇个人局域网之间的短距离数据交换（使用 2.4～2.485GHz 的 ISM 波段的 UHF 无线电波）。蓝牙可连接多个设备，克服了数据同步的难题。蓝牙技术联盟负责监督蓝牙标准的制定，项目的管理、认证，并维护商标权益。

master-slave　　主从（模式）。这里指在蓝牙通信时，必须有一个主端（master），以及其他从端（slave）。由主端进行查找，发起配对，配对成功后方可收发数据。理论上，一个主端可同时与 7 个从端进行通信。

embedded device　　嵌入式设备，指带有微型计算机（一般是单片机）的小型设备，主要由嵌入式处理器、相关支撑的硬件和软件系统组成，它是集软、硬件于一体的可独立工作的智能设备。

RFID　　*abbr.* 无线射频识别（Radio Frequency IDentification），也称电子标签，是一种通信技术，可通过无线电信号识别特定目标并读写数据，无须在识别系统与识别对象之间建立机械或光学接触。

smartcard　　智能卡，是内嵌微芯片的塑料卡的通称，其大小通常与一张信用卡相当，配备有 CPU 和 RAM，一些智能卡还配有 RFID 芯片。

CSMA/CD　　*abbr.* 带冲突检测的载波监听多路访问（Carrier Sense Multiple Access with Collision Detection），是共享信道时的通信规则，即以太网协议。

frame　　帧，是数据链路层的信息封装单位，网络层、传输层对应的单位则分别是数据包（packet）和报文（segment）。

WLAN　　*abbr.* 无线局域网（Wireless Local Area Network），利用射频（Radio Frequency）技术用无线电波取代以线缆构成的局域网。

IEEE 802.11　　无线局域网的通用标准，由国际电气和电子工程师协会（IEEE）定义。

ad hoc mode　　ad hoc 模式，是一种省去了无线中介设备而搭建起来的对等网络结构。

WEP　　*abbr.* 有线等效加密（Wired Equivalent Privacy），是对两台设备间通过无线传输的数据进行加密的方式，用以防止非法用户窃听和侵入无线网络。

WPA　　*abbr.* Wi-Fi 保护接入（Wi-Fi Protected Access），是一种保护无线数据网络安全的系统，在克服 WEP 的几个严重弱点的基础上产生。

microsecond　　微秒，1×10^{-6} 秒，常写为 μs。此外常用的还有 millisecond（毫秒，ms，1×10^{-3} 秒），nanosecond（纳秒，ns，1×10^{-9} 秒），picosecond（皮秒，ps，1×10^{-12} 秒），femtosecond（飞秒，fs，1×10^{-15} 秒）。

Mbps　　兆比特每秒。此外还有 Kbps，即千比特每秒；Gbps，即吉比特每秒。

100BASE-TX　　这是快速以太网三种与传输介质有关的标准之一，另外两种是 100BASE-T4 和 100BASE-FX。

Difficult Sentences Analysis and Translation 难句分析与翻译

1. LANs are restricted in size, which means that the worst-case transmission time is bounded and known in advance.

译文：局域网的规模是受限制的，这意味着在最糟糕情况下的传输时间有上限，且可以预知。

分析：句中短语 "be restricted in..." 意思是 "在……方面受限制"，短语 "in advance" 的意思是 "提前、预先"；which 引导非限制性定语从句，which 指代前面整个主句；"that the worst-case transmission time is bounded and known in advance" 是宾语从句。

2. A Wireless Local Area Network (WLAN) is a wireless computer network that links two or more devices using a wireless distribution method within a limited area such as a home, school, computer laboratory, or office building.

译文：无线局域网（WLAN）是一种使用无线分布式方法连接两台或两台以上设备的无线计算机网络，在诸如家庭、学校、计算机实验室或者办公大楼等有限区域内使用。

分析：句中 that 引导定语从句，修饰、限定前面的 "a wireless computer network"；现在分词短语 "using a wireless distribution method" 表示方式。

3. The Wide Area Network technology is a method of creating an intelligent data sharing network of computers, whose complexity and scale is so huge that it may extend over a city, a region, or a country.

译文：广域网技术是一种共享计算机网络来生成智能数据的方法，其复杂性和规模之巨大，以至于可以扩展到一个城市、一个地区或一个国家。

分析：句中 "so...that..." 引导结果状语从句；"whose complexity and scale is..." 是非限制性定语从句，翻译时可以单独成句。

Module 3　Consolidation Exercise 巩固练习

K2.PDF

I. Answer the following questions according to the texts

1. What is one important criterion for classifying networks?

2. What do you know about the first version of Ethernet?

3. Why were switches introduced to replace hubs in Ethernet-based networks?

4. What standards are most modern WLANs based on?

5. What basic modes of operation does the IEEE 802.11 have?

6. Why did the 802.11 designers also include encryption mechanisms like Wired Equivalent Privacy and Wi-Fi Protected Access?

7. What causes a blurring of the distinction between LANs and WANs?

8. Why do private organizations like big software companies have their own WANs constructed nowadays?

II. Give the meaning of the following abbreviated terms both in English and in Chinese

PAN	CSMA/CD
MAN	WPA
UTP	LAN
WLAN	RFID
WEP	WAN

III. Translate the following sentences into Chinese

1. To help these users, some companies got together to design a short-range wireless network called Bluetooth to connect these components without wires.

2. Knowing this bound makes it possible to use certain kinds of designs that would not otherwise be possible.

3. Ethernet was designed to accommodate multiple computers that were interconnected on a shared bus topology.

4. A significant development that enhanced LAN performance was the introduction of switches to replace hubs in Ethernet-based networks.

5. WLANs have become popular in the home due to ease of installation and use, and in commercial complexes offering wireless access to their customers, often for free.

6. In infrastructure mode, mobile units communicate through an access point that serves as a bridge to other networks (such as Internet or LAN).

7. Many access points will also offer Wi-Fi Protected Setup, a quick (but now insecure) method of joining a new device to an encrypted network.

8. All data that is transmitted on the Internet is always labeled with the destination address which is a numerical name assigned to every computer connected on the network.

Lesson 3　PSTN, GSM and WWAN

Module 1　Text Study 课文学习

Basic Training 基本训练

Text 1　根据语音和视频完成以下任务。

Task 1-0 听录音，记录关键词，理解课文大意。

L3-1-1.MP3

L3-1-2.MP3

____

____

____

____

____

____

Task 1-1 翻译词汇或句子中的画线部分。

L3-1.MP4

公共电话交换网______________________________

移动电话________　简称为________　固定电话________　固定线路____________

由于移动电话的智能化，移动电话也称智能电话____________，由于移动电话使用蜂窝网络，所以也称蜂窝电话____________。

模拟的________　数字的________　电路交换____________　分组交换____________

无缝地________　地址空间____________　层次__________　扁平的____________

采样率____________　解析率__________　编码____________　本地回路__________

脉冲编码调制______________________________

The PSTN consists of __________（电话线）, ________（光缆）,__________（微波传输链路）, _______（蜂窝网络）, ____________（通信卫星）and ____________（海底电话电缆）.

The technical operation of the PSTN follows（或 ad…______）the standards created by the ITU-T.

To carry a typical __________（电话呼叫）from a __________（主叫方）to a __________（被叫方）, the analog audio signal is digitized.

Task 1-2 词汇辨析。

digital 和 digitalized _____。

Task 1-3 用英语回答将电话交换设计为层次结构的原因。

_____

_____

Task 1-4 翻译 "The trunks connecting the exchanges are also digital, called circuits or channels." 理解 trunk、circuit 和 channel 的语义。

_____

_____

Task 1-5 翻译 "The analog audio signal is digitized at an 8 kHz sample rate with 8-bit resolution using a special type of nonlinear Pulse Code Modulation (PCM) known as G.711." 理解 sample rate、resolution（ratio）中 rate 和 ratio 的语义。此句常用专业术语较集中，建议默写和背诵，以达到能熟练运用的目的。

_____

_____

Task 1-6 填写图（Figure 1-3）中各物的英文名称。

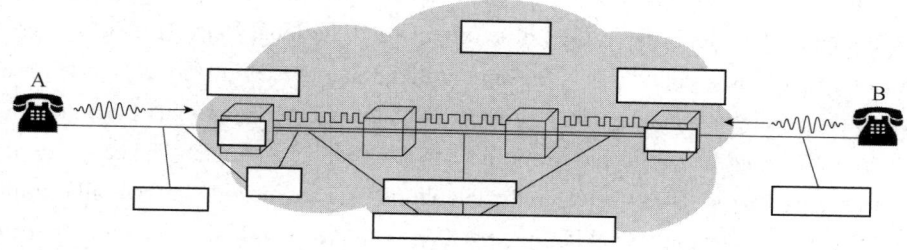

Figure 1-3   Cloze Test

课文

PSTN

The Public Switched Telephone Network (PSTN) (Figure 1-4) is the aggregate of the world's circuit-switched telephone networks that are operated by national, regional, or local telephone operators, providing infrastructure and services for public telecommunication. The PSTN consists of telephone lines, fiber optic cables, microwave transmission links, cellular networks, communications satellites, and undersea telephone cables, all interconnected by switching centers, thus allowing most telephones to communicate with each other. Originally a network of fixed-line analog telephone systems, the PSTN is now almost entirely digitalized in its core network and includes mobile and other networks, as well as fixed telephones.

The technical operation of the PSTN adheres to the standards created by the ITU-T. These standards allow different networks in different countries to interconnect seamlessly. The E.163 and E.164 standards provide a single global address space for telephone numbers. The combination of the interconnected networks and the single numbering plan allows telephones around the world to dial each other.

Figure 1-4　PSTN and Cellular Networks

The original concept was that the telephone exchanges are arranged into hierarchies, so that if a call cannot be handled in a local cluster, it is passed to one higher up for onward routing. This reduced the number of connecting trunks required between operators over long distances and also kept local traffic separate.

However, in modern networks the cost of transmission and equipment is lower and, although hierarchies still exist, they are much flatter, with perhaps only two layers.

As described above, most automated telephone exchanges now use digital switching rather than mechanical or analog switching. The trunks connecting the exchanges are also digital, called circuits or channels. However, analog (Figure 1-5) two-wire circuits are still used to connect the last mile from the exchange to the telephone in the home (also called the local loop). To carry a typical phone call from a calling party to a called party, the analog audio signal is digitized at an 8 kHz sample rate with 8-bit resolution using a special type of nonlinear Pulse Code Modulation (PCM) known as G.711. The call is then transmitted from one end to another via telephone exchanges. The call is switched using a call set up protocol (usually ISUP) between the telephone exchanges under an overall routing strategy.

The call is carried over the PSTN using a 64 Kbps channel, originally designed by Bell Labs.

telephone

PSTN network

cellular network

Figure 1-5　Analog Switching

Text 2　根据语音和视频完成以下任务。

Task 2-0 听录音，记录关键词，理解课文大意。

L3-2-1.MP3

L3-2-2.MP3

_____

_____

_____

Task 2-1 GSM 是_____的缩写，也是一个商标，
把缩写作为商标的例子还有_____。

Task 2-2 比较 institute、association 和 organization 的词义，例如：
_____翻译为"协会"，是较为正式的组织，强调兴趣和需要的一致性；
_____指科学、教育的机构、研究所、学会；_____指（较通用的）组织、机构。

Task 2-3 develop、research、study、deploy、explore、implement、deliver 这些动词出现在
一项工程的不同阶段，从工程顺序的角度对它们进行排序。

_____

Task 2-4 5G 的含义可能是_____，也可能是_____。

Task 2-5 transmit、transfer 和 transport 都可以翻译为_____，它们在语义上的差别是：
_____。

Task 2-6 modem 和 codec 分别是_____和_____、_____和_____的合成词。

Task 2-7 理解 capacity 的词义。
在日常生活中指人的_____，容器的_____；在 IT 领域，可以是网络的
_____，内存的_____，CPU 的_____。

Task 2-8 词汇练习。
"集成"的名词形式是_____，动词形式是_____；"集成电路"的全称和缩写
是_____。

Task 2-9 GPRS 的英文全称是什么？

_____

Task 2-10 词汇翻译。

国家和地区_____ 超过 90%的市场份额_____

基站_____或_____ 互联网接入_____

语音电话_____

Task 2-11 "as of 2014"等于"by 2014"，表示"到 2014"，复习和回顾叙述发展过程的
词和词组：as of、over time、first、then、subsequently。

Task 2-12 填空。

It takes _____ of the _____ in mobile technology and the _____ of mobile
infrastructure.

课文

GSM，Cellular Network and Wireless WAN

GSM (Global System for Mobile Communications) is a standard developed by the European
Telecommunications Standards Institute (ETSI) to describe the protocols for second-generation (2G)
digital cellular networks used by mobile phones. As of 2014 it has become the default global standard
for mobile communications—with over 90% market share, operating in over 219 countries and
territories. 2G networks developed as a replacement for first generation (1G) analog cellular networks,

and the GSM standard originally described a digital, circuit-switched network optimized for full duplex voice telephony. This expanded over time to include data communications, first by circuit-switched transport, then by packet data transport via GPRS (General Packet Radio Services) and EDGE (Enhanced Data rates for GSM Evolution or EGPRS). Subsequently, the 3GPP developed third-generation (3G) UMTS standards followed by fourth-generation (4G) LTE Advanced standards, which do not form part of the ETSI GSM standard. "GSM" is a trademark owned by the GSM Association. It may also refer to the most common voice codec used, Full Rate.

A cellular network or mobile network is a communication network where the last link is wireless. The network is distributed over land areas called cells, each served by at least one fixed-location transceiver, known as a cell site or base station. This base station provides the cell with the network coverage which can be used for transmission of voice, data and others. In a cellular network, each cell uses a different set of frequencies from neighboring cells, to avoid interference and provide guaranteed bandwidth within each cell.

When joined together, these cells provide radio coverage over a wide geographic area. This enables a large number of portable transceivers (e.g. mobile phones, pagers, etc.) to communicate with each other and with fixed transceivers and telephones anywhere in the network, via base stations, even if some of the transceivers are moving through more than one cell during transmission.

Cellular networks offer a number of desirable features:

● More capacity than a single large transmitter, since the same frequency can be used for multiple links as long as they are in different cells.

● Mobile devices use less power than with a single transmitter or satellite since the cell towers are closer.

● Larger coverage area than a single terrestrial transmitter, since additional cell towers can be added indefinitely and are not limited by the horizon.

Major telecommunications providers have deployed voice and data cellular networks over most of the inhabited land area of the earth. This allows mobile phones and mobile computing devices to be connected to the public switched telephone network and public Internet. Private cellular networks can be used for research or for large organizations and fleets, such as dispatch for local public safety agencies or a taxicab company.

WAN technology has also taken the step to the next level of network integration which is based on the mobile phone platform. It provides Internet connectivity through the wireless telecommunication mode and is called Wireless Wide Area Network (WWAN). It is a part of the third generation (3G) mobile technologies that are being offered. It takes advantage of the advances in mobile technology and the development of mobile infrastructure to deliver high quality and high-speed Internet access. The routing mechanisms and protocols for this new technology are different.

Advanced Training 进阶训练

Task 读懂课文，并根据提示恰当地填写（或补足）所需的英语词汇。

T3.PDF

5G

In telecommunications, 5G (Figure 1-6) is the _____（第五代）technology standard for _____（蜂窝网络）, which cellular phone companies began deploying worldwide in 2019, the planned successor to the 4G networks which provide _____（连接性）to most current cellphones. Like its predecessors, 5G networks are cellular networks, in which the service area is divided into small geographical areas called cells. All 5G wireless devices in a cell are connected to the Internet and telephone network by _____（无线电波）through a local antenna in the cell. The main _____（进步）of the new networks is that they will have greater _____（带宽）, giving higher download speeds, eventually _____（高达）10 gigabits per second (Gbit/s). Due to the increased bandwidth, it is expected that the new networks will not just serve _____（蜂窝电话）like existing cellular networks, but also be used as _____（一般的）service providers for _____（笔记本电脑…tops）and _____（台式计算机 desk…）computers, competing with existing _____（互联网服务商 Ps）such as cable Internet, and also will make possible new applications in Internet of Things (IoT) and machine to machine areas. Current 4G cellphones will not be able to use the new networks, which will require new 5G enabled wireless devices.

Figure 1-6　GPP's 5G logo

Vocabulary Practice 词汇练习

词汇及短语听写，并纠正发音。

C3.MP3

_____

_____

_____

Words & Expressions

aggregate [ˈægrɪɡət]	n. 聚集；合计；总计
adhere [ədˈhɪə(r)]	vi. 遵循；附着
analog [ˈænəlɔːɡ]	adj.（计算机）模拟的；（钟表）有长、短针的 n. 类似物；（计算机）模拟
seamlessly [ˈsiːmlɪslɪ]	adv. 无空隙地；无停顿地
hierarchies [ˈhaɪərɑːkɪz]	n.（hierarchy 的复数形式）层次体系
resolution [ˌrezəˈluːʃn]	n. 分辨率，解析度；决议
pager [ˈpeɪdʒə]	n. 寻呼机
transceiver [trænˈsiːvə(r)]	n. 无线电收发机、收发器；收发报机
terrestrial [təˈrestrɪəl]	adj. 陆地的；地球的；人间的
dispatch [dɪˈspætʃ]	n. 快速处理；急件；派遣

Module 2　Study Aid 辅助帮学

Terms & Abbreviations 术语和缩写

circuit-switched　　　电路交换。"交换"就是按照某种方式动态地分配传输信道资源，分为电路交换（circuit-switched/circuit-switching）和分组交换（packet-switching）。电路交换在通信期间独占物理信道。分组交换可将用户传输的数据划分成一定长度的分组，由交换机根据每个分组的地址标志转发至目的地。

PSTN   *abbr.* 公共交换电话网络（Public Switched Telephone Network）是一种旧式电话系统，即我们日常生活中的固定电话网。

ITU-T   *abbr.* 国际电信联盟电信标准化部门（Telecommunication Standardization Sector of ITU）是国际电信联盟管理下的专门制定电信标准的分支机构。

sample rate   采样频率，指模拟信号在进行数字化时的单位时间采样次数。

PCM   *abbr.* 脉冲编码调制（Pulse Code Modulation），是对连续变化的模拟信号进行抽样、量化和编码以产生数字信号的过程。

GSM   *abbr.* 全球移动通信系统（Global System for Mobile Communications），是当前应用最广泛的移动电话系统。

GPRS   *abbr.* 通用分组无线服务（General Packet Radio Services），是 GSM 移动电话的一种移动数据业务，属于第二代移动通信中的数据传输服务。

3GPP   *abbr.* 第三代合作伙伴计划（3rd Generation Partnership Project），目标是实现由 2G 网络到 3G 网络的平滑过渡，保证未来技术的后向兼容性，支持轻松建网及系统间的漫游和兼容。3GPP 的职能主要以制定 GSM 核心网为基础，UTRA 为无线接口的第三代技术规范。

UMTS   *abbr.* 通用移动通信系统（Universal Mobile Telecommunications System），是一个完整的 3G 移动通信系统，但并不仅限于定义空中接口。

LTE   *abbr.* 长期演进（Long Term Evolution），是指由第三代合作伙伴计划组织制定的通用移动通信系统技术标准的长期演进，于 2004 年 12 月在 3GPP 多伦多会议上正式立项并启动。

ETSI   *abbr.* 欧洲电信标准化协会（European Telecommunications Standards Institute），是由欧共体委员会于 1988 年批准并建立的一个非营利性的电信标准化组织，总部设在法国南部的尼斯。

cellular network   蜂窝网络，是移动网络（mobile network）的通信硬件架构，把移动电话的服务区分为一个个正六边形的小子区，每个小子区设一个基站，形成了形状酷似"蜂窝"的结构。

WWAN   *abbr.* 无线广域网（Wireless Wide Area Network），该网络使笔记本电脑或者其他设备在蜂窝网络覆盖范围内以无线方式连接到互联网。

Difficult Sentences Analysis and Translation 难句分析与翻译

1. The PSTN consists of telephone lines, fiber optic cables, microwave transmission links, cellular networks, communications satellites and undersea telephone cables, all interconnected by switching centers, thus allowing most telephones to communicate with each other.

译文：公共交换电话网络包括电话线、光纤线缆、微波传输链路、蜂窝网络、通信卫星和海底电话电缆，全都通过交换中心相互连接，这样大多数电话可以相互通信。

分析：句中短语"consists of"意思是"包括"；"all interconnected by switching centers"是分词独立结构，作状语，表示伴随方式；现在分词短语"thus allowing most telephones to communicate with each other"作状语，表示结果。

2. The original concept was that the telephone exchanges are arranged into hierarchies, so that if a call cannot be handled in a local cluster, it is passed to one higher up for onward routing.

译文：最初的概念是电话交换机按照层次结构安排，所以如果一个呼叫不能在本地集群中处理，那么它就会向上传递到更高层次的路由。

分析：句中 "that the telephone exchanges are arranged into hierarchies" 是表语从句；"so that…" 引导的目的状语从句中还有一个条件状语从句 "if a call cannot be handled in a local cluster"。

3. To carry a typical phone call from a calling party to a called party, the analog audio signal is digitized at an 8 kHz sample rate with 8-bit resolution using a special type of nonlinear Pulse Code Modulation (PCM) known as G.711.

译文：为传输典型的从呼叫方到被叫方的电话呼叫，通过采用一种被称为 G.711 的特殊类型的非线性脉冲编码调制（PCM），模拟音频信号以 8 比特的分辨率和 8 kHz 的采样率进行数字化。

分析：句中不定式短语 "To carry a typical phone call…" 是目的状语；现在分词短语 "using a special type of nonlinear Pulse Code Modulation (PCM) known as G.711" 是方式状语。

4. This enables a large number of portable transceivers (e.g. mobile phones, pagers, etc.) to communicate with each other and with fixed transceivers and telephones anywhere in the network, via base stations, even if some of the transceivers are moving through more than one cell during transmission.

译文：这使得大量便携式收发器（例如，移动电话、寻呼机等）可以在网络中的任何地方经由基站相互通信，并与固定的收发器及电话通信，即使在传输期间有些收发器移动经过了不止一个蜂窝单元。

分析：句中短语 "enables… to …" 意思是 "使……能够……"；"even if some of the transceivers are moving through more than one cell during transmission" 是让步状语从句。

Module 3   Consolidation Exercise 巩固练习

K3.PDF

I. Translate the following terms into English

采样频率_____   蜂窝网络_____

无线广域网_____   寻呼机_____

公共交换电话网络_____   脉冲编码调制_____

全球移动通信系统_____   通用移动通信系统_____

II. Read the following statements and decide whether they are true or false according to the passage. Put "T" for True and "F" for False

1. Telephones around the world can dial each other based on the combination of the interconnected networks and the single numbering plan.

2. In modern networks the cost of transmission and equipment is a lot higher because hierarchies still exist with perhaps at least three layers.

3. The call is switched using a call set up protocol (usually ISUP) between the telephone

exchanges originally designed by Bell Labs.

4. GSM is a standard developed by the European Telecommunications Standards Institute to describe the protocols for 2G digital cellular networks used by mobile phones.

5. One of the desirable features cellular networks offer is that mobile devices use more power than with a single transmitter or satellite since the cell towers are closer.

6. Major telecommunications providers have deployed voice and data cellular networks over most of the uninhabited land area of the earth.

7. WAN technology has also taken the step to the next level of network integration based on the mobile phone platform.

8. The routing mechanisms and protocols for WWAN technology remain the same as those for WAN.

III. Translate the following sentences into Chinese

1. The PSTN is now almost entirely digitalized in its core network and includes mobile and other networks, as well as fixed telephones.

2. The technical operation of the PSTN adheres to the standards created by the ITU-T.

3. As described above, most automated telephone exchanges now use digital switching rather than mechanical or analog switching.

4. The call is then transmitted from one end to another via telephone exchanges.

5. The GSM standard originally described a digital, circuit-switched network optimized for full duplex voice telephony.

6. A cellular network or mobile network is a communication network where the last link is wireless.

7. Major telecommunications providers have deployed voice and data cellular networks over most of the inhabited land area of the earth.

8. It takes advantage of the advances in mobile technology and the development of mobile infrastructure to deliver high quality and high-speed Internet access.

Lesson 4 Examples of Network Application

Module 1 Text Study 课文学习

Basic Training 基本训练

Text 1 根据语音和视频完成以下任务。

Task 1-0 听录音，记录关键词，理解课文大意。

_____

_____ L4-1-1.MP3

_____

_____

_____ L4-1-2.MP3

_____

_____

_____ L4-1.MP4

Task 1-1 写出并区分描述通信过程的一些相似、相反的英语词汇。

发送_____ 传递_____ 接收_____

接受（确认、承认）_____ 投递、交付_____

分发_____ 存、取_____ 检索、恢复、取回_____

个人的_____ 职业的_____ 私人的_____

公有的_____ 请求_____ 响应_____

回复_____ 客户端_____ 服务端_____

"客户端/服务端"的英文缩写_____

Task 1-2 receiver 和 recipient 的区别是_____。

Task 1-3 用中文解释 host 的含义。

host 在日常生活中指_____；对一台计算机而言是指_____，以区分于外围设备（peripheral equipment）；在互联网中，host 指_____；在人工智能中，host 指_____。host 用于表示一种具有_____（性质）的东西。

Task 1-4 翻译画线部分。

网关_____是有协议_____转换能力的网络设备，也可以泛指与其他网络连接的中转设备，尤其是路由器_____和防火墙_____。

Task 1-5 根据提示填空。

① Businesses have experienced a rapid _____（增加）in productivity and _____（减少）in expenses, and found new ways to _____（提升）their services.

② POP and IMAP (Internet Message _____（存取）Protocol) are the two most prevalent Internet standard protocols for E-mail _____（检索）。

Task 1-6 写出下列缩写的英文全称。

POP3_____   SMTP_____

Task 1-7 词汇辨析。

① 名词词性的 connection 和 link_____。

② Server 和 service_____。

③ Message 和 information_____。

课文

E-mail

E-mail (Electronic-mail, Figure 1-7) is a technology that includes passing and sending information from one place to another, using a computer and the Internet. It has proven beneficial in our personal as well as professional life. A majority of businesses around the globe use E-mail as the most employed method of rapid and effective office communication. It is an important tool for corporate communication. From the time E-mail has been used as a communication tool, businesses have experienced a rapid rise in productivity and reduction in expenses, and found new ways to improve their services. The advantages of E-mail communication can be seen in large-scale as well as small-scale industries and organizations. However, the advantages are not limited only to the business world. We also use E-mails in our day-to-day lives for sending and receiving messages from our friends and family members.

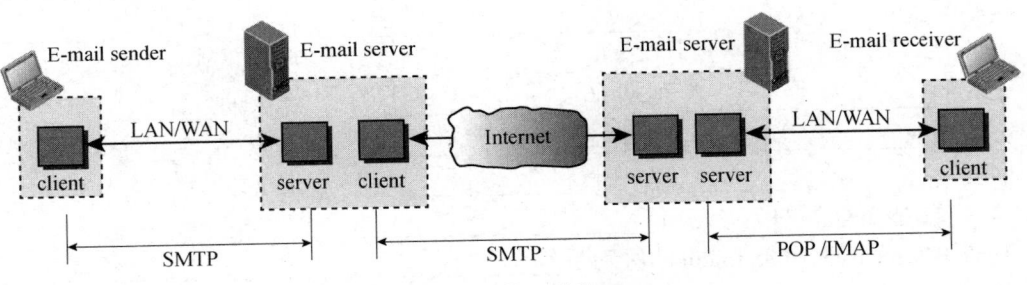

Figure 1-7   E-mail System

By using TCP, SMTP (Simple Mail Transfer Protocol) sends E-mail to other computers that support the TCP/IP protocol suite. SMTP provides extension to the local mail services that existed in the early years of LANs. It supervises the E-mail sending from the local mail host to a remote mail host. It is not reliable for accepting mail from local users or distributing received mail to recipients. This is the responsibility of the local mail system.

SMTP makes use of TCP to establish a connection to the remote mail host, the mail is sent, any waiting mail is requested and then the connection is disconnected. It can also return a forwarding address if the intended recipient no longer receives E-mail at that destination. To enable mail to be delivered across differing systems, a mail gateway is used.

In computing, the Post Office Protocol (POP) is an application layer Internet standard protocol used by local E-mail clients to retrieve E-mail from a remote server over a TCP/IP connection. POP and IMAP (Internet Message Access Protocol) are the two most prevalent Internet standard protocols for E-mail retrieval. Virtually all modern E-mail clients and servers support both. The POP protocol has been developed through several versions, with version 3 (POP3) being the current standard. Like IMAP, POP3 is supported by most webmail services such as Hotmail, Gmail and Yahoo Mail.

Text 2　根据语音和视频完成以下任务。

Task 2-0 听录音，记录关键词，理解课文大意。

_____

_____　L4-2-1.MP3

_____

_____

_____　L4-2-2.MP3

_____

Task 2-1 区分相反意思的词。

上传_____ 下载_____　L4-2.MP4

明文_____ 密文_____

Task 2-2 写出常使用的信息单位的英文。

比特_____ 字节_____ 字_____ 帧_____

包_____ 会话_____ 文件_____ 文件夹_____

Task 2-3 manipulate 和 operate 都是"操作、处理"的意思，但_____还有"控制"的意思，可以翻译成"操纵、操控"。

Task 2-4 翻译。

与……无关_____ 文件管理器_____

浏览器_____

Task 2-5 -oriented 和-based 都是表示"以……为基础"的语言结构，但-oriented 有较固定的意思，connection-oriented 翻译为_____，类似的结构还有 Web-enabled（支持 Web）。

Task 2-6 翻译。

User data is interspersed in-band with Telnet control information in an 8-bit byte-oriented data connection over the Transmission Control Protocol (TCP).

_____

Task 2-7 RFC 是_____的缩写。

Task 2-8 根据提示填写英文词汇。

FTP is that it can be considered as one of the best and fastest means _____（或 w…或 m…）of file transfer over a computer network.

Task 2-9 Command-Line Interface 的缩写是_____; GUI 是_____的缩写。

Task 2-10 "Because of security issues with Telnet, its use for this purpose has waned in favor of SSH." 句中的 "security issues" 可以翻译成 "安全问题"，它和 security problems 有区别吗？

Task 2-11 写出主要的中文词义。

interactive_____   intercept_____

intersperse_____   interlink_____

Task 2-12 "通过……协议" 通常使用介词_____。

Task 2-13 数据链路层的地址是_____，而网络层和传输层的则是_____和_____。

课文

FTP

FTP stands for File Transfer Protocol. It is a method to transfer files over the Internet from one computer to another. It provides you an easy way to upload and download files between computers that are connected to the Internet.

FTP is a network protocol that is used to transfer data between computers over a network. It uses the ports 20 and 21. FTP is used for the exchange of files over a TCP network. It uses separate connections for control and data. Port 21 is used for control while port 20 is used for data. FTP is a client-server protocol in which an FTP client connects to an FTP server to access and manipulate the files on the server.

The File Transfer Protocol is used to send files from computers hosting Web servers. It can be used for downloading files from servers and for sending files from one computer to another. One of the primary advantages of FTP is that it can be considered as one of the best and fastest means of file transfer over a computer network. It is an efficient way of transferring large files among computers in a network. It is popularly used for exchanging files between computers irrespective of the operating systems they run. Most Web browsers and file managers are capable of connecting to FTP servers, whereby the clients can upload and download files to and from Web servers.

Telnet

Telnet is a client-server protocol, based on a reliable connection-oriented transport. Typically, this protocol is used to establish a connection to Transmission Control Protocol (TCP) port number 23, where a Telnet server application is listening. Telnet, however, predates TCP/IP and was originally run over Network Control Program (NCP) protocols.

Telnet is a network protocol used on the Internet or Local Area Networks to provide a bidirectional interactive text-oriented communications facility using a virtual terminal connection. User data is interspersed in-band with Telnet control information in an 8-bit byte-oriented data connection over the Transmission Control Protocol (TCP).

Telnet was developed in 1969, beginning with RFC 15, extended in RFC 854, and standardized as Internet Engineering Task Force (IETF) Internet Standard STD 8, one of the first Internet standards.

Historically, Telnet provided access to a command-line interface (usually, of an operating system) on a remote host. Most network equipment and operating systems with a TCP/IP stack support a Telnet service for remote configuration (including systems based on Windows NT). Because of security issues with Telnet, its use for this purpose has waned in favor of SSH (Secure SHell).

While the Telnet protocol supports user authentication, it does not support the transport of encrypted data. All data exchanged during a Telnet session is transported as plain text across the network. This means that the data can be intercepted and easily understood.

If security is a concern, SSH protocol offers an alternate and secure method for server access. SSH provides a structure for secure remote login and other secure network services. It also provides stronger authentication than Telnet and supports the transport of session data using encryption. As a best practice, network professionals should always use SSH in place of Telnet whenever possible.

Advanced Training 进阶训练

T4.PDF

Task 根据提示填写英文。

WWW

The World Wide Web is a _____（基于超文本的 hypertext-…d）system for finding and _____（存取 ac…）Internet resources, and it is a set of programs, standards, and protocols governing the way in which _____（多媒体…media）files are created and displayed on the Internet. It can provide access to a variety of Internet resources from the same _____（界面 in…）, including FTP, Gopher. The World Wide Web is a distributed, multimedia and hypertext system, and thus a unique medium for communications and for publishing.

The term WWW is used to describe an _____（相互连接…linked）system of documents in which a user may jump from one document to another in a nonlinear, associative way. The ability to jump from one document to the next is made possible via the use of _____（超链接…links）. By clicking on the hyperlink, the user is immediately connected to the document specified by the link. Most documents on the Web are written in HTML.

HTTP _____（代表 s…）for Hypertext Transfer Protocol. It is a communication protocol that facilitates the transfer of information on the Internet. It is a _____（请求—响应 r…-r…）protocol between clients and servers. Clients are Web users or Web browsers, while the responding server that stores or creates the resources requested is known as the origin server. HTTP can be implemented on any of the Internet protocols. However, the TCP/IP protocol suite is most popularly used. An HTTP client establishes a TCP connection with the host. Port 80 is the default _____（端口）used for the establishment of this connection between the client and the server using HTTP. On receiving the request, the server _____（回复…）with a status line, a message and the requested resource. Since a protocol is set of rules and procedures for communication on a network and since HTTP is a protocol—HTTP in itself is a set of rules and procedures used to communicate over the World Wide Web.

World Wide Web Consortium (W3C) and the Internet Engineering Task Force (IETF) coordinated HTTP development. Its initial purpose was to provide a way to retrieve and publish

HTML documents. HTTP protocol comes under _____（应用层…）(Layer 5) of the TCP/IP model. HTTP clients make requests to the HTTP protocol and the concerned HTTP server handles these requests. The clients making requests to the Web server are referred to as user agents. Typical clients are Web browsers, search engine spiders, _____（Web-…支持的）applications, etc.

Vocabulary Practice 词汇练习

词汇及短语听写，并纠正发音。

C4.MP3

_____

_____

_____

_____

Words & Expressions

beneficial [ˌbenɪ'fɪʃl]	*adj.* 有利的，有益的
productivity [ˌprɒdʌk'tɪvətɪ]	*n.* 生产率
extension [ɪk'stenʃn]	*n.* 伸展，扩大；延长，延期
supervise ['su:pəvaɪz]	*v.* 监督，管理
recipient [rɪ'sɪpɪənt]	*n.* 收件人
retrieve [rɪ'tri:v]	*v.* 取回；检索
prevalent ['prevələnt]	*adj.* 流行的
irrespective [ˌɪrɪ'spektɪv]	*adj.* 无关的，不考虑的
predate ['pri:'deɪt]	*v.* 居先；（在日期上）早于，先于
intersperse [ˌɪntə'spɜ:s]	*v.* 点缀；散布，散置
wane [weɪn]	*v.* 衰落；结束
intercept [ˌɪntə'sept]	*v.* 拦截
nonlinear [nɒn'lɪnɪə(r)]	*adj.* 非线性的
default [dɪ'fɔ:lt]	*adj.* 缺省的，默认的

Module 2　Study Aid 辅助帮学

Terms & Abbreviations 术语和缩写

E-mail　*abbr.* 电子邮件（Electronic-mail），一种用电子手段提供信息交换的通信方式，是互联网最早和最为广泛的应用之一。

gateway　　网关，又称网间连接器、协议转换器，在网络层以上实现网络互联，是最复杂的网络互联设备，能互联两个高层协议不同的网络。

POP　*abbr.* 邮局协议（Post Office Protocol），用于电子邮件的接收。

IMAP　*abbr.* 互联网信息访问协议（Internet Message Access Protocol），以前称作交互邮件访问协议（Interactive Mail Access Protocol）。IMAP 是斯坦福大学在 1986 年开发的一种邮件获取协议，它的主要作用是让邮件客户端（如 MS Outlook Express）从邮件服务器上获取邮件的信息、下载邮件等。它与 POP3 协议的主要区别是用户可以不用下载全部邮件，而在客户端上直接对服务器上的邮件进行操作。

connection-oriented   面向连接。通信双方在通信时，要事先建立一条通信线路，包括建立连接、使用连接和释放连接三个过程。建立连接需要分配相应的资源，如缓冲区，以保证通信能正常进行。

NCP   *abbr.* 网络控制程序（Network Control Program）。

CLI   *abbr.* 命令行界面（Command-Line Interface），也称 Text-Oriented Interface，是在图形用户界面（GUI）得到普及之前使用最广泛的用户界面，通常不支持鼠标，用户通过键盘输入指令，屏幕上通常也只显示文字，也有人称为字符用户界面（CUI）。

RFC   *abbr.* 请求评估（Request for Comments），是一系列以编号排定的文件，文件收集了有关互联网、UNIX 系统的信息和互联网社区的软件文件。RFC 文件还额外加入了许多讨论议题，例如，互联网新开发的协议及过程中所有的记录。基本的互联网通信协议都在 RFC 文件内有详细说明，几乎所有的互联网标准都收录在 RFC 文件中。目前 RFC 文件是由 Internet SOCiety（ISOC）赞助、发行的。

SSH   *abbr.* 安全外壳（Secure SHell）协议，由 IETF 的网络工作小组制定，是建立在应用层和传输层上的安全协议。这是目前较可靠的、专为远程登录会话和其他网络服务提供安全的协议，最初是 UNIX 系统上的一个程序，后来迅速扩展到了其他操作平台上。

Gopher   Gopher 是 Internet 上一个非常有名的信息查找系统，它将 Internet 上的文件组织成某种索引，很方便地将用户从 Internet 的一处带到另一处。在 WWW 出现之前，Gopher 是 Internet 上最主要的信息检索工具，Gopher 站点也是最主要的站点。但在 WWW 出现后，Gopher 基本过时，人们很少再使用它。

HTML   *abbr.* 超文本标记语言（Hyper Text Markup Language），是标准通用标记语言下的一个应用，也是一种规范和标准，它通过标记符号来定义要显示在网页中的各个部分。

W3C   *abbr.* 万维网联盟（World Wide Web Consortium），又称 W3C 理事会，是国际上最著名的标准化组织之一。在该联盟成立后，至今已发布近百项万维网相关的标准，对万维网发展做出了杰出的贡献。

Difficult Sentences Analysis and Translation 难句分析与翻译

1. E-mail (Electronic-mail) is a technology that includes passing and sending information from one place to another, using a computer and the Internet.

译文：电子邮件是一种通过一台计算机和互联网，提供从一个地方到另一个地方传递信息的技术。

分析：句中"that includes passing and sending information…"是定语从句，修饰、限定其前的名词"technology"；动名词短语"passing and sending information from one place to another"作动词"includes"的宾语；现在分词短语"using a computer and the Internet"表示方式。

2. Telnet is a network protocol used on the Internet or Local Area Networks to provide a bidirectional interactive text-oriented communications facility using a virtual terminal connection.

译文：Telnet 是一种在互联网或者局域网上使用的网络协议。该协议使用虚拟终端连接来提供面向文本的双向交互式通信功能。

分析：句中过去分词短语"used on the Internet or Local Area Networks"作定语，修饰、限定"a network protocol"；动词不定式短语"to provide a bidirectional interactive text-oriented communications facility"表示目的；现在分词短语"using a virtual terminal connection"表示方式。

3. The World Wide Web is a hypertext-based system for finding and accessing Internet resources, and it is a set of programs, standards, and protocols governing the way in which multimedia files are created and displayed on the Internet.

译文：万维网是一个基于超文本的用于查找和访问互联网资源的系统。它是一组管理多媒体文件在互联网上创建及显示方式的程序、标准和协议。

分析：句中动名词短语"finding and accessing Internet resources"作介词宾语；现在分词短语"governing the way…"作定语，修饰、限定其前面的名词"programs, standards, and protocols"；"in which multimedia files are created and displayed on the Internet"是定语从句。

Module 3　Consolidation Exercise 巩固练习

K4.PDF

I. Translate the following phrases into Chinese, and give the established abbreviations

Network Application＿＿＿＿＿＿＿＿＿＿＿＿＿＿＿＿＿＿＿＿＿＿

Simple Mail Transfer Protocol＿＿＿＿＿＿＿＿＿＿＿＿＿＿＿＿＿

Post Office Protocol＿＿＿＿＿＿＿＿＿＿＿＿＿＿＿＿＿＿＿＿＿＿

Internet Message Access Protocol＿＿＿＿＿＿＿＿＿＿＿＿＿＿＿

File Transfer Protocol＿＿＿＿＿＿＿＿＿＿＿＿＿＿＿＿＿＿＿＿＿

Web browsers＿＿＿＿＿＿＿＿＿＿＿＿＿＿＿＿＿＿＿＿＿＿＿＿＿

Network Control Program＿＿＿＿＿＿＿＿＿＿＿＿＿＿＿＿＿＿＿

Transmission Control Protocol＿＿＿＿＿＿＿＿＿＿＿＿＿＿＿＿＿

Command-Line Interface＿＿＿＿＿＿＿＿＿＿＿＿＿＿＿＿＿＿＿＿

Hypertext Transfer Protocol＿＿＿＿＿＿＿＿＿＿＿＿＿＿＿＿＿＿

Secure SHell＿＿＿＿＿＿＿＿＿＿＿＿＿＿＿＿＿＿＿＿＿＿＿＿＿

Hypertext Markup Language＿＿＿＿＿＿＿＿＿＿＿＿＿＿＿＿＿＿

II. Fill in each blank with the information from the texts

1. From the time E-mail has been used as a communication tool, businesses have experienced a rapid rise in＿＿＿＿＿and＿＿＿＿＿in expenses.

2. SMTP provides＿＿＿＿＿to the local mail services that existed in the early years of LANs, and it＿＿＿＿＿the E-mail sending from the local mail host to a remote mail host.

3. One of the primary＿＿＿＿＿of FTP is that it can be considered as one of the best and fastest means of file＿＿＿＿＿over a computer network.

4. Telnet is a network protocol used on the Internet or Local Area Networks to provide a＿＿＿＿＿interactive text-oriented communications facility using＿＿＿＿＿＿＿＿＿.

5. While the Telnet protocol supports user＿＿＿＿＿＿＿, it does not support the transport of

_____data.

6. The World Wide Web is a set of programs, standards, and_____governing the way in which multimedia files are_____on the Internet.

7. The term WWW is used to describe an interlinked system of_____in which a user may jump from one document to another in_____.

8. The initial purpose of HTTP development was to provide a way to_____and_____HTML documents.

III. Translate the following sentences into Chinese

1. A majority of businesses around the globe use E-mail as the most employed method of rapid and effective office communication.

2. The mail can also return a forwarding address if the intended recipient no longer receives E-mail at that destination.

3. FTP provides you an easy way to upload and download files between computers that are connected to the Internet.

4. Typically, this protocol is used to establish a connection to Transmission Control Protocol port number 23, where a Telnet server application is listening.

5. Historically, Telnet provided access to a command-line interface on a remote host.

6. As a best practice, network professionals should always use SSH in place of Telnet, whenever possible.

7. By clicking on the hyperlink, the user is immediately connected to the document specified by the link.

8. Clients are Web users or Web browsers, while the responding server that stores or creates the resources requested is known as the origin server.

Lesson 5 Internet of Things and Electronic Commerce

Module 1 Text Study 课文学习

Basic Training 基本训练

Text 1 根据语音和视频完成以下任务。

Task 1-0 听录音，记录关键词，理解课文大意。

_____ L5-1-1.MP3

_____

_____

_____ L5-1-2.MP3

_____

_____

_____ L5-1.MP4

Task 1-1 翻译画线部分。

<u>物联网</u>_____，又称传感网，指的是将各种信号传感设备，如<u>射频识别</u>_____装置、<u>红外感应器</u>_____、<u>全球定位系统</u>_____、<u>激光扫描器</u>_____、<u>执行器</u>_____等可以<u>相互操作</u>_____的装置与互联网结合起来而形成的一个巨大的<u>信息物理系统</u>_____，包括_____一些技术，例如，<u>智能电网</u>_____，其目的是让<u>所有的物与物</u>_____都通过网络连接在一起，方便<u>识别</u>_____ 和<u>管理</u>_____。

Task 1-2 用于泛指事物、东西、事项等的词有_____、_____和_____。

Task 1-3 various、a variety of 和 a wide variety of 翻译为_____。

Task 1-4 <u>自主的</u>_____和<u>自动的</u>_____这两个词的词义很接近，从哲学的角度来看，没有本质上的区别，只是在技术上"自主的"自动化程度更高一些，有"自主、自治"能力，能更大程度脱离人的干涉。

Task 1-5 embedded 的同义词是_____。

Task 1-6 flow 表示"流动"时，是及物动词还是不及物动词？近义词是_____。

Task 1-7 和中文一样，英文也可以用_____符号表示有并列关系的几个词或句子。

Task 1-8 翻译。

机器人就是增加了传感器和执行器的计算机。

_____

Task 1-9 写出英文全名。

M2M_____ Wi-Fi_____

Task 1-10 翻译及比较。

① "Things" in the sense of IoT can refer to a wide variety of devices with built-in sensors.

② "Things" from the perspective of IoT can refer to a wide variety of devices with embedded sensors.

Task 1-11 Internet of Things 和 Things of Internet 在语义上有区别吗？

Task 1-12 举例说明 instance 和 example 的语义区别。

课文

Internet of Things

The Internet of Things (IoT) is the network of physical objects, such as devices, vehicles, buildings and other items which are embedded with electronics, software, sensors and network connectivity, which enables these objects to collect and exchange data. The Internet of Things allows objects to be sensed and controlled remotely across existing network infrastructure, creating opportunities for more-direct integration between the physical world and computer-based systems, and resulting in improved efficiency, accuracy and economic benefit. When IoT is augmented with sensors and actuators, the technology becomes an instance of the more general class of Cyber-Physical Systems (CPS), which also encompasses technologies such as smart grids, smart homes, intelligent transportation and smart cities. Each thing is uniquely identifiable through its embedded computing system, but is able to interoperate within the existing Internet infrastructure.

British entrepreneur Kevin Ashton first coined the term in 1999 while working at Auto-ID Labs (originally called Auto-ID centers)，referring to a global network of Radio Frequency IDentification (RFID) connected objects. Typically, IoT is expected to offer advanced connectivity of devices, systems, and services that goes beyond Machine-to-Machine communications (M2M) and covers a variety of protocols, domains and applications. The interconnection of these embedded devices (including smart objects), is expected to usher in automation in nearly all fields, while also enabling advanced applications like a smart grid, and expanding to the areas such as smart cities.

"Things" in the sense of IoT, can refer to a wide variety of devices such as heart monitoring implants, biochip transponders on farm animals, electric clams in coastal waters, automobiles with built-in sensors, DNA analysis devices for environmental/food/pathogen monitoring or field operation devices that assist fire fighters in search and rescue operations. These devices collect useful data with the help of various existing technologies and then autonomously flow the data between other devices. Current market examples include smart thermostat systems and washer/dryers that use Wi-Fi for remote monitoring.

Text 2 根据语音和视频完成以下任务。

Task 2-0 听录音，记录关键词，理解课文大意。

L5-2-1.MP3

_____

_____

_____

_____

L5-2-2.MP3

L5-2.MP4

Task 2-1 翻译。

Besides the plethora of new application areas for Internet-connected automation to expand into, IoT is also expected to generate large amounts of data from diverse locations that is aggregated very quickly, thereby increasing the need to better index, store and process such data.

_____

_____

句中 "that is aggregated very quickly" 是_____从句，修饰、限定先行词_____，作为结果状语的现在分词短语是_____。

Task 2-2 填写或翻译相关的中/英文词汇。

① 课文中 surge 的近义词是_____。

② 快速响应码_____是光学标签_____的一种，具有超高速识读的特点。

③ 雾计算_____是云计算_____的延伸概念，由思科（Cisco）首创。这个因"云"而"雾"的命名源自"雾是更贴近地面的云"这条名句。

④ 近场通信_____由 RFID 发展而来，也可以看作 RFID 的子集，RFID 一般只具有单向_____数据传输能力，而近场通信具有双向_____通信和计算能力。

⑤ Internet of Things requires huge _____（可扩展性）in the network space to _____（应对）the _____（涌入）of devices.

⑥ RFID was the _____ technology. Later, NFC became dominant.

⑦ 轻量数据传输_____ 蓝牙低能耗_____ 纽扣电池_____
实时可扩展性_____ 以节能的方式_____ 可行的替代物_____

课文

New Technologies

Besides the plethora of new application areas for Internet-connected automation to expand into, IoT is also expected to generate large amounts of data from diverse locations that is aggregated very quickly, thereby increasing the need to better index, store and process such data. IoT is one of the platforms of today's Smart City and Smart Energy Management Systems.

Internet of Things (Figure 1-8) requires huge scalability in the network space to handle the surge of devices. IETF（Internet Engineering Task Force）6LoWPAN would be used to connect devices to IP networks. With billions of devices being added to the Internet space, IPv6 will play a major role in handling the network layer scalability. IETF's Constrained Application Protocol, MQTT and ZeroMQ would provide lightweight data transport.

Fog computing is a viable alternative to prevent such large burst of data flow through Internet. The edge devices' computation power can be used to analyze and process data, thus providing easy real-time scalability.

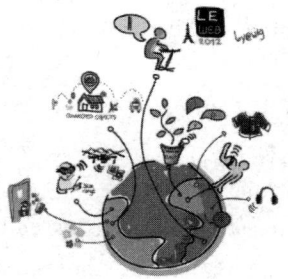

Figure 1-8　Internet of Things

There are mainly three types of technologies that enable IoT:

- RFID and NFC. In the 2000s, RFID was the dominant technology. Later, NFC became dominant. NFC has become common in smartphones during the early 2010s, with uses such as reading NFC tags or for access to public transportation.

- Optical tags and quick response codes. This is used for low cost tagging. A phone camera decodes QR code using image-processing techniques. In reality QR advertisement campaigns give less turnout as users need to have another application to read QR codes.

- Bluetooth Low Energy. This is one of the latest technologies. All newly releasing smartphones have BLE hardware in them. Tags based on BLE can signal their presence at a power budget that enables them to operate for up to one year on a lithium coin cell battery.

Advanced Training 进阶训练

Task 根据提示补全词汇，并阅读理解原文。

T5.PDF

Electronic Commerce

Electronic commerce, _____ （通常 c...ly） written as E-commerce, is the trading or facilitation of trading in products or services using computer networks, such as the Internet. Electronic commerce _____[draws on 利用（可替换为 makes..._____of）] technologies such as mobile commerce, electronic funds transfer, supply chain management, Internet marketing, _____（在线交易） processing, Electronic Data Interchange (EDI), inventory management systems, and automated data collection systems. Modern electronic commerce typically uses the World Wide Web for at least one part of the transaction's life cycle, although it may also use other technologies such as E-mail.

E-commerce businesses may employ some or all of the following:

- Online shopping Web sites for retail sales direct to consumers.

Online shopping Web is a form of electronic commerce which allows consumers to directly buy goods or services from a seller over the Internet using a Web browser. Alternative names are: e-Web-store, e-shop, e-store, Internet shop, Web-shop, Web-store, online store, online storefront and virtual store. _____ （移动商务）(or M-commerce) describes purchasing from an online retailer's mobile _____（优化 op...ed） online site or app.

● Providing or participating in online marketplaces, which process third-party. …-to-c…（B2C）or …-to-…（C2C）sales.

Online marketplace is a type of E-commerce site where product or service information is provided by multiple third parties, whereas _____（交易）are processed by the marketplace operator. Online marketplaces are the primary type of _____（多渠道 m…cha…l）ecommerce. In an online marketplace, consumer transactions are processed by the marketplace operator and then delivered and fulfilled by the participating retailers or wholesalers (often called drop shipping). Other capabilities might include auctioning (forward or reverse), catalogs, ordering, wanted advertisement, trading exchange functionality and capabilities like RFQ, RFI or RFP.

Vocabulary Practice 词汇练习

C5.MP3

词汇及短语听写，并纠正发音。

_____

_____

_____

_____

Words & Expressions

integration [ˌɪntɪˈɡreɪʃn]	n. 整合；一体化；结合
augment [ɔːɡˈment]	v. 增强，加强；增加
actuator [ˈæktʃʊeɪtə]	n. [电脑]执行机构、执行器；激励者
encompass [ɪnˈkʌmpəs]	v. 围绕、包围；包含、包括
identifiable [aɪˌdentɪˈfaɪəbl]	adj. 可辨认的、可识别的；可视为相同的
interoperate [ˌɪntərˈɒpəreɪt]	v. 交互操作
coin [kɔɪn]	v. 创造；杜撰
usher [ˈʌʃə(r)]	v. 引领
implant [ɪmˈplænt]	n.（植入身体中的）移植物
biochip [ˈbaɪətʃɪp]	n. 生物芯片
transponder [trænsˈpɒndə(r)]	n. 发射机应答器，询问机，转发器
pathogen [ˈpæθədʒən]	n. 病菌，病原体
thermostat [ˈθɜːməstæt]	n. 恒温（调节）器
plethora [ˈpleθərə]	n. 过多，过剩
surge [sɜːdʒ]	n. 激增；急剧上升；汹涌
scalability [skeɪləˈbɪlɪtɪ]	n. 可扩展性
transaction [trænˈzækʃn]	n. 交易，业务，事务
inventory [ˈɪnvəntrɪ]	n. 存货清单；财产目录；清查
retail [ˈriːteɪl]	v. 零售；零卖 n. 零售业 adj. 零售的
optimized [ˈɒptɪmaɪzd]	adj. 最佳化的，（使）最优化的
wholesaler [ˈhəʊlseɪlə(r)]	n. 批发商
auction [ˈɔːkʃn]	n. 拍卖；竞卖 vt. 拍卖；竞卖
accountancy [əˈkaʊntənsɪ]	n. 会计职业，会计工作；会计学

audit [ˈɔːdɪt]　　　　　　　　　　　　*n.* 审计，查账　*v.* 审计
demographic [ˌdeməˈɡræfɪk]　　　*adj.* 人口统计学的；人口统计的
newsletter [ˈnuːzˌletə]　　　　　　 *n.* 内部通信

Module 2　Study Aid 辅助帮学

Terms & Abbreviations　术语和缩写

CPS　　　*abbr.* 信息物理系统（Cyber-Physical Systems），是集计算、通信与控制于一体的新一代智能系统，是计算进程和物理进程的统一体。该系统通过人机交互接口实现和物理进程的交互，使用网络化空间以远程的、可靠的、实时的、安全的、协作的方式操控一个物理实体。

6LoWPAN　　　6LoWPAN 是 IPv6 over Low power Wireless Personal Area Networks 的简写，是一种基于 IPv6 的低速无线个域网标准，即 IPv6 over IEEE 802.15.4。

ZeroMQ　　　ZeroMQ 简称 ZMQ，是一个简单、好用的传输层，像框架一样的 Socket Library，使得 Socket 编程更加简单、简洁和高性能。

MQTT　　　*abbr.* 消息队列遥测传输（Message Queuing Telemetry Transport），是 IBM 开发的一个即时通信协议，有可能成为物联网的重要组成部分。该协议支持所有平台，几乎可以把所有联网物品和外部连接起来，被用作传感器和控制设备的通信协议。

NFC　　　*abbr.* 近场通信（Near-Field Communication），一种短距、高频无线通信技术，以 13.56MHz 的频率运行于 20 厘米的距离内，其传输速度有 106Kbps、212Kbps 和 424Kbps 三种。

BLE　　　*abbr.* 蓝牙低能耗（Bluetooth Low Energy），是一种低成本、短距离、可互操作的鲁棒性无线技术，工作在免许可的 2.4GHz ISM 射频频段，能利用许多智能手段最大限度地降低功耗。

E-commerce　　　*abbr.* 电子商务（electronic commerce），指在互联网和增值网（Value Added Network，VAN）上以电子方式进行交易活动和相关服务活动，是传统商业活动各环节的电子化和网络化。

EDI　　　*abbr.* 电子数据交换（Electronic Data Interchange），是由国际标准化组织（ISO）推出、使用的国际标准，指一种为商业或行政事务处理，按照一个公认的标准，形成结构化的事务处理或消息报文格式，从计算机到计算机的电子传输方法，也是计算机可识别的商业语言。例如，国际贸易中的采购订单、装箱单、提货单等数据的交换。

M-commerce　　　*abbr.* 移动电子商务（mobile commerce），就是利用手机、掌上电脑等无线终端进行的 B2B、B2C、C2C 或 ABC 的电子商务。它将互联网、移动通信技术、短距离通信技术及其他信息处理技术完美结合，使人们可以在任何时间和地点进行各种商贸活动，实现随时随地、线上线下的购物与交易、在线电子支付，以及各种交易活动、商务活动、金融活动和相关的综合服务活动等。

drop shipping　　　转运配送或直接代发货，是供应链管理中的一种方法。零售商不需商品库存，而是把客户订单和装运细节给批发商，供货商将货物直接发送给最终客户。零售商赚取批发价格和零售价格之间的差价。

online transaction　　　在线事务处理，在有关商务的地方可以翻译成网上交易。

RFQ　　*abbr.* 报价请求（Request for Quotation），在外贸函电中作为买方给卖方发一个询盘，可能采用电子邮件格式或传真格式，传真文件的表头或者邮件的主题就会出现"RFQ"的字样。

RFI　　*abbr.* 信息请求（Request for Information），是一个标准的业务流程，其目的是收集关于各种供应商能力的书面信息，通常遵循一种可用于比较的格式。

RFP　　*abbr.* 征求建议书（Request for Proposal），是发单人在向数家承包商征求解决方案建议时，向外招标并发放的一种文件。类似于 RFQ，即要求报价，不过比 RFQ 复杂得多，通常需要提供一份建议书。

Difficult Sentences Analysis and Translation 难句分析与翻译

1. Besides the plethora of new application areas for Internet-connected automation to expand into, IoT is also expected to generate large amounts of data from diverse locations that is aggregated very quickly, thereby increasing the need to better index, store and process such data.

参考译文：连接了自动化的物联网，为互联网扩展了大批新的应用领域，除此之外，它也必然从不同的位置快速汇集并生成大量的数据，因此，增加了对更好地索引、存储和处理这类数据的需求。

分析：句中"that is aggregated very quickly"是定语从句，修饰、限定先行词"data"，现在分词短语"increasing the need to better index…"作结果状语。

2. The Internet of Things (IoT) is the network of physical objects, such as devices, vehicles, buildings and other items which are embedded with electronics, software, sensors and network connectivity, which enables these objects to collect and exchange data.

译文：物联网（IoT）是由物理对象（如设备、车辆、建筑物和嵌入了电子器件、软件、传感器和网络连接的其他装置）组成的网络。网络内的所有对象能互相收集和交换数据。

分析：句中定语从句"which are embedded with electronics, software, sensors and network connectivity"修饰、限定前面的"other items"；非限制性定语从句"which enables these objects to collect and exchange data"中的"which"指代"the network"。

3. When IoT is augmented with sensors and actuators, the technology becomes an instance of the more general class of Cyber-Physical Systems (CPS), which also encompasses technologies such as smart grids, smart homes, intelligent transportation and smart cities.

译文：物联网增加传感器和执行器后，该技术便成为更为普遍的信息物理系统的一个实例，且包括智能电网、智能家居、智能交通和智慧城市等技术。

分析：句中"When IoT is augmented with sensors and actuators"作时间状语；非限制性定语从句"which also encompasses technologies…"中的"which"指代"the technology"。

4. Online marketplace is a type of E-commerce site where product or service information is provided by multiple third parties, whereas transactions are processed by the marketplace operator.

译文：在线市场是一种电子商务网站，这里的产品或服务信息由多个第三方提供，而交易则由市场运营商处理。

分析：句中"where product or service information is provided by multiple third parties"是定语从句，修饰、限定前面的"E-commerce site"；"whereas"是连词，此处表示对比，译为"而"。

Module 3   Consolidation Exercise 巩固练习

I. Answer the following questions according to the passage

K5.PDF

1. How was the term IoT first made?

2. Two current market examples of IoT that use Wi-Fi for remote monitoring are mentioned in Text 1, what are they?

3. Why will IPv6 play a major role in handling the network layer scalability?

4. What are the three main types of technologies that enable IoT?

5. What is electronic commerce?

6. What do you know about online shopping Web?

II. Give the meaning of the following abbreviated terms both in English and in Chinese

RFI	RFQ	EDI	QR code
IoT	CPS	E-commerce	M-commerce
BLE	RFP		

III. Translate the following sentences into Chinese

1. Each thing is uniquely identifiable through its embedded computing system, but is able to interoperate within the existing Internet infrastructure.

2. These devices collect useful data with the help of various existing technologies and then

autonomously flow the data between other devices.

3. Internet of Things requires huge scalability in the network space to handle the surge of devices.

4. The edge devices' computation power can be used to analyze and process data, thus providing easy real-time scalability.

5. NFC has become common in smartphones during the early 2010s, with uses such as reading NFC tags or for access to public transportation.

6. Tags based on BLE can signal their presence at a power budget that enables them to operate for up to one year on a lithium coin cell battery.

7. Mobile commerce describes purchasing from an online retailer's mobile optimized online site or app.

8. In an online marketplace, consumer transactions are processed by the marketplace operator and then delivered and fulfilled by the participating retailers or wholesalers.

Lesson 6   Network Storage and Cloud Computing

Module 1   Text Study 课文学习

Basic Training 基本训练

Text 1   根据语音和视频完成以下任务。

Task 1-0 听录音，记录关键词，理解课文大意。

L6-1.MP3

_____

_____

Task 1-1 翻译词汇和句子中的画线部分。   L6-1.MP4

clustered file system_____ distributed file system_____

file hosting service_____   逻辑池_____   logical space_____

本地协同_____，类似的语言结构还有合作_____。

Task 1-2 比较下列词汇和短语的细微区别。

utilize_____   make use of_____   take advantage of_____

employ_____   adopt_____   use_____

Task 1-3 根据提示填空，注意词性和分词用法。

① These cloud storage providers are responsible for keeping the data _____ （有效的）and _____ （可访问的）, and the physical environment _____ （受保护的）and _____ （运行）.

② People buy or lease _____ （存储容量）from the providers to store user data.

③ API is shorted for_____.

④ NAS stands for_____.

⑤ The meaning of SAN is_____.

课文

The Overview of Network Storage

Network storage may refer to:

- Cloud storage.
- Clustered file system or distributed file system.
- File hosting service.
- File server.
- Network-Attached Storage (NAS).
- Storage Area Network (SAN).

Cloud storage is a model of data storage in which the digital data is stored in logical pools. The physical storage spans multiple servers (and often locations), and the physical environment is typically owned and managed by a hosting company. These cloud storage providers are responsible for keeping the data available and accessible, and the physical environment protected and running. People and organizations buy or lease storage capacity from the providers to store users, organization, or application data. Cloud storage services may be accessed through a co-located cloud computer service, a Web service Application Programming Interface (API) or by applications that utilize the API, such as cloud desktop storage, a cloud storage gateway or Web-based content management systems.

Text 2   根据语音和视频完成以下任务。

Task 2-0 听录音，记录关键词，理解课文大意。   L6-2-1.MP3

_____

_____

_____

_____

_____

_____   L6-2-2.MP3

_____

_____

_____   L6-2-3.MP3

_____

_____

_____   L6-2-4.MP3

Task 2-1 翻译词汇和句子中的画线部分。

加载_____   与位置无关_____

highlight_____   client–server scheme_____   L6-2.MP4

simultaneous_____和 synchronous_____语义不同，在通信中没有共同的时钟时为异步的_____。

workstation 是使用服务器_____提供服务_____的计算机，即客户机_____，也指大（中、小）型机的终端_____，也指性能很好的个人计算机_____，例如图形工作站_____。

可靠性_____   复杂性_____   冗余性_____

存储_____   检索、取回_____   网盘_____

中等规模_____   自我包含_____

apply_____   application_____   appliance_____

APP_____   API_____

Task 2-2 "There are several approaches to clustering." 中的 approach 在这里翻译成_____，同 way、method 和 means 等的意思一样，approach 起源于 ap（去）+proach_____。

Task 2-3 address、host 作为动词时的词义分别是_____和_____。

Task 2-4 找出课文中表示"执行、从事"意思的词和词组。

_____

Task 2-5 翻译和比较。

① It does not run programs for the benefit of its clients._____

② It does not run programs on behalf of its clients._____

Task 2-6 根据专业技术背景，写出对应的英语词汇。

① 文件级别_____、块级别_____或 block-based，都是指数据的抽象程度（level of abstraction）和上层应用（upper application）能使用（employ）的数据结构（data structure）。

② NAS is _____（专门化）for serving files either by its hardware, software, or _____（配置）.

Task 2-7 built for a specific purpose 可以简单地表达为_____。

Task 2-8 根据课文，用英文回答问题并解释原因：Is removable HDD a DAS or NAS?

_____

_____

Task 2-9 翻译。

① 这是一个为特定目的建造的应用设备。

_____

② 不必要将 DAS 网络化。

_____

③ SAN leaves file system concerns on the "client" side.

_____

Task 2-10 翻译，并比较词性和词义。

① The config zone you configured contains your configuration.

_____

② config_____ configure_____ configuration_____

③ compute_____ computer_____ computation_____ computational_____

课文

The Details of Network Storage

A clustered file system is a file system which is shared by being simultaneously mounted on multiple servers. There are several approaches to clustering, most of which do not employ a clustered file system (only Direct-Attached Storage for each node). Clustered file systems can provide features like location-independent addressing and redundancy which improve reliability or reduce the complexity of the other parts of the cluster. Parallel file systems are a type of clustered file system that spread data across multiple storage nodes, usually for redundancy or performance.

A file hosting service, cloud storage service, online file storage provider, or cyberlocker is an Internet hosting service specifically designed to host user files. It allows users to upload files that could then be accessed over the Internet from a different computer, tablet, smart phone or other networked device, by the same user or possibly by other users, after a password or other authentication is provided. Typically, the services allow HTTP access, and sometimes FTP access. Related services are content-displaying hosting services (i.e., video and image), virtual storage, and remote backup. In computing, a file server (or fileserver) is a computer attached to a network that has the primary purpose of providing a location for shared disk access, i.e., shared storage of computer files (such as documents, sound files, photographs, movies, images, databases, etc.) that can be accessed by the workstations that are attached to the same computer network. The term server highlights the role of the machine in the client-server scheme, where the clients are the workstations using the storage. A file server is not intended to perform computational tasks, and does not run programs on behalf of its clients. It is designed primarily to enable the storage and retrieval of data while the computation is carried out by the workstations.

File servers are commonly found in schools and offices, where users use a LAN to connect their client computers.

Network-Attached Storage (NAS) is a file-level computer data storage server connected to a computer network providing data access to a heterogeneous group of clients. NAS is specialized for serving files either by its hardware, software, or configuration. It is often manufactured as a computer appliance—a purpose-built specialized computer. NAS systems are networked appliances which contain one or more storage drives, often arranged into logical, redundant storage containers or RAID (Figure 1-9). The key difference between Direct-Attached Storage (DAS) and NAS is that DAS is simply an extension to an existing server and is not necessarily networked(Figure 1-10). NAS is designed as an easy and self-contained solution for sharing files over the network. Both DAS and NAS can potentially increase availability of data by using RAID or clustering. NAS (NAS Network is shown in Figure 1-11) provides both storage and a file system. This is often contrasted with SAN (Storage Area Network), which provides only block-based storage and leaves file system concerns on the "client" side. SAN protocols include Fibre Channel, internet Small Computer System interface (iSCSI), ATA over Ethernet (AoE) and HyperSCSI.

Figure 1-9   RAID

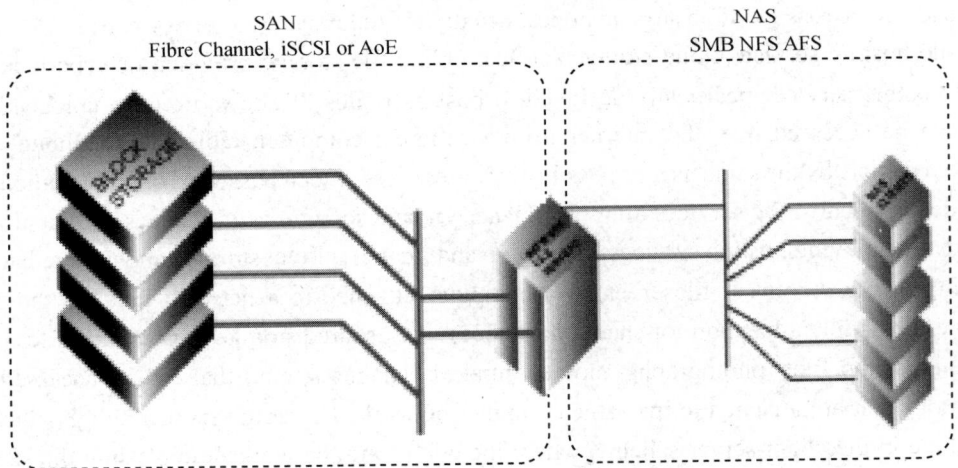

Figure 1-10   SAN & NAS

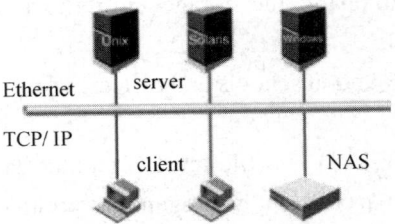

Figure 1-11   NAS Network

A Storage Area Network (SAN) is a network which provides access to consolidated, block level data storage. SANs are primarily used to enhance storage devices, such as disk arrays, tape libraries, and optical jukeboxes, accessible to servers so that the devices appear to the operating system as locally attached devices. A SAN typically has its own network of storage devices that are generally not accessible through the Local Area Network (LAN) by other devices. The cost and complexity of SANs dropped in the early 2000s to levels allowing wider adoption across both enterprise and small to medium-sized business environments. A SAN does not provide file abstraction, only block-level operations. However, file systems built on top of SANs do provide file-level access, and are known as shared-disk file systems.

Advanced Training 进阶训练

Cloud Computing

T6.PDF

Cloud computing is a relatively new term that conveys the use of information technology services and resources that are provided on a service basis. According to a 2008 IEEE paper, "Cloud computing is a paradigm in which information is permanently stored in servers on the Internet and cached temporarily on clients that include desktops, entertainment centers, table computers, notebooks, wall computers, sensors, monitors, etc."

In network diagrams, resources that are provided by an outside entity are depicted in a "cloud" formation. In the current (but still evolving) model of cloud computing, the cloud computing infrastructure consists of services that are offered up and delivered through data centers that can be accessed from anywhere in the world. The cloud in this model is the single point of access for the computing needs of the customers being serviced.

In the cloud computing definitions that are evolving, the services in the cloud are being provided by enterprises and accessed by others via the Internet. The resources are accessed in this manner as a service—often on a subscription basis. Users of the services being offered often have very little knowledge of the technology being used. The users also have no control over the infrastructure that supports the technology they are using.

In cloud computing models, customers do not own the infrastructure they are using, and they basically rent it, or pay as they use it. The loss of control is seen as a negative, but it is generally outweighed by several positives. One of the major selling points of cloud computing is lower costs. Companies will have lower technology—based capital expenditures, which should enable companies to focus their money on delivering the goods and services that they specialize in. There will be more device and location independence, enabling users to access systems no matter where they are located or what kind of device they are using. The sharing of costs and resources amongst so many users will also allow for efficiencies and cost savings around things like performance, load balancing, and even locations (locating data centers and infrastructure in areas with lower real estate costs, for example). Cloud computing is also thought to affect reliability and scalability in positive ways. One of the major topics in information technology today is data security. In a cloud infrastructure, security typically improves overall, although there are concerns about the loss of control over some sensitive data.

Finally, cloud computing results in improved resource utilization, which is good for the sustainability movement (i.e., green technology or clean technology).

There is a big push for cloud computing services by several big companies. Amazon.com has been at the forefront of the cloud computing movement. Google and Microsoft have also been very publicly working on cloud computing offerings. Some of the other companies to watch for in this field are Yahoo, IBM, Intel, HP and SAP. Several large universities have also been busy with large scale cloud computing research projects.

Task 1 参考上文，将下列词组或语句片段放在一个简短、完整的语句中。

① …term that conveys the use of…

_____

② …on a service basis…

_____

③ …a paradigm in which…

_____

④ …permanently…temporarily…

_____

⑤ …diagrams, …are depicted…

_____

⑥ …negative, …positives

_____

Task 2 在讨论云计算的利弊时，课文中用到了一些与计算性能相关的词和词组，尝试进行翻译。

资源利用_____   控制的失去_____
较低的开销_____   负载均衡_____
可靠性_____   可扩展性_____
数据安全_____   敏感数据_____

Vocabulary Practice 词汇练习

词汇及短语听写，并纠正发音。   C6.MP3

_____

_____

Words & Expressions

lease [li:s]   *n.* 租约 *vt.* 出租
simultaneous [ˌsɪmlˈteɪnɪəs]   *adj.* 同时的，同步的
redundancy [rɪˈdʌndənsɪ]   *n.* 冗余度
highlight [ˈhaɪlaɪt]   *vt.* 强调，突出
heterogeneous [ˌhetərəˈdʒi:nɪəs]   *adj.* 各种各样的
appliance [əˈplaɪəns]   *n.* 装置
consolidate [kənˈsɒlɪdeɪt]   *v.* 合成一体

jukebox [ˈdʒuːkbɔks]	*n.* 自动唱片点唱机，光盘盒
convey [kənˈveɪ]	*vt.* 表达
paradigm [ˈpærədaɪm]	*n.* 范例，样式
permanently [ˈpɜːmənəntlɪ]	*adv.* 永久地，长期不变地
depict [dɪˈpɪkt]	*v.* 描绘
formation [fɔːˈmeɪʃn]	*n.* 结构，形状
subscription [səbˈskrɪpʃn]	*n.* 会员费，捐助
outweigh [ˌaʊtˈweɪ]	*v.* （在重要性或价值方面）超过
sustainability [səˌsteɪnəˈbɪlətɪ]	*n.* 持续性，能维持性，永续性
forefront [ˈfɔːfrʌnt]	*n.* 最前部；活动中心；前列；第一线

Module 2　Study Aid 辅助帮学

Terms & Abbreviations　术语和缩写

network storage　　网络存储，是数据存储的一种方式，大致分为三种结构：直连式存储、网络附加存储和存储区域网络。

cloud computing　　云计算，是基于互联网的相关服务的增加、使用和交付模式，通常涉及通过互联网来提供动态、易扩展且经常是虚拟化的 CPU 资源。

cloud storage　　云存储，是在云计算概念上延伸和发展出来的一个新的概念，是一种新兴的网络存储技术，是指通过集群应用、网络技术或分布式文件系统等，将网络中大量各种不同类型的存储设备通过应用软件集合起来协同工作，共同对外提供数据存储和业务访问功能的一个系统。

clustered file system　　集群文件系统，是指运行在多台计算机之上和之间，通过某种方式相互通信从而将集群内所有存储空间资源整合和虚拟化，并对外提供文件访问服务的文件系统。

distributed file system　　分布式文件系统，该文件系统管理的物理存储资源不一定直接连接在本地节点上，而是通过计算机网络与节点相连的。分布式文件系统的设计基于客户机/服务器模式。

NAS　　*abbr.* 网络附加存储（Network-Attached Storage），是一种采用直接与网络介质相连的特殊设备，实现数据存储的机制。由于设备都分配有 IP 地址，所以客户机通过充当数据网关的服务器可以对其进行存取、访问，甚至在某些情况下，不需要任何中间介质客户机也可以直接访问这些设备。

SAN　　*abbr.* 存储区域网络（Storage Area Network），是指存储设备相互连接且与一台或多台服务器相连的网络，将服务器用作其接入点。在有些配置中也与网络相连。SAN 将特殊交换机当作连接设备。它们看起来很像常规的以太网络交换机，是 SAN 中的连通点。SAN 使得在各自网络上实现相互通信成为可能，同时带来了很多有利条件。

API　　*abbr.* 应用程序编程接口（Application Programming Interface），是一些预先定义的函数，目的是提供应用程序与开发人员基于某软件或硬件得以访问一组例程的能力，而又无须访问源码，或理解内部工作机制的细节。

cyberlocker   网络存储平台，又名网络空间、上传空间，简称网空，是云端服务的一种形式，在安全存储架构中提供远程存储空间的在线数据托管服务，可以通过 Internet 在全球访问。

RAID   *abbr.* 磁盘阵列（Redundant Array of Independent Disks），有"独立磁盘构成的具有冗余能力的阵列"之意。磁盘阵列是由很多价格较便宜的磁盘组合成一个容量巨大的磁盘组，利用不同磁盘同时提供数据来提升磁盘系统整体效能。利用这项技术，可以将数据切割成许多区段，分别存放在各个硬盘上。

DAS   *abbr.* 直连式存储（Direct-Attached Storage），是一种直接与主机系统连接的存储设备，目前仍是计算机系统中最常用的数据存储方法。

wall computer   墙上的巨屏计算机。

Difficult Sentences Analysis and Translation 难句分析与翻译

1. Cloud storage is a model of data storage in which the digital data is stored in logical pools, the physical storage spans multiple servers (and often locations), and the physical environment is typically owned and managed by a hosting company.

译文：云存储是数据存储的一种模型。在该模型中，数字数据存储在逻辑池中，物理存储跨越多个服务器（通常是在本地），物理环境通常由托管公司拥有和管理。

分析：句中"in which the digital data is stored … by a hosting company"是定语从句，修饰、限定"a model of data storage"；在该定语从句中，"the digital data is stored in logical pools""the physical storage spans multiple servers (and often locations)"和"the physical environment is typically owned and managed by a hosting company"是并列的分句。

2. Cloud storage services may be accessed through a co-located cloud computer service, a Web service Application Programming Interface (API) or by applications that utilize the API, such as cloud desktop storage, a cloud storage gateway or Web-based content management systems.

译文：云存储服务可以通过协同定位的云计算机服务、Web 服务应用程序编程接口或由利用该接口的这类应用访问，诸如云桌面存储、云存储网关或者基于 Web 的内容管理系统。

分析：句中"that utilize the API"是定语从句，修饰、限定先行词"applications"；在该定语从句中，"such as"引出的是"applications"的实例。

3. In computing, a file server (or fileserver) is a computer attached to a network that has the primary purpose of providing a location for shared disk access, i.e., shared storage of computer files (such as documents, sound files, photographs, movies, images, databases, etc.) that can be accessed by the workstations that are attached to the same computer network.

译文：在计算中，文件服务器就是连接到网络的计算机，而该网络的主要目的在于提供共享磁盘访问的位置，即共享存储计算机文件（例如文档、声音文件、照片、电影、图像、数据库等）。连接到同一计算机网络的工作站都可以访问这些文件。

分析：句中过去分词短语"attached to a network"作先行词"computer"的定语；"that has the primary purpose of providing a location for shared disk access"是定语从句，修饰、限定先行词"network"；定语从句"that can be accessed by the workstations"修饰、限定先行词"computer files"；"that are attached to the same computer network"是定语从句，修饰、限定先行词"workstations"。

4. Cloud computing is a relatively new term that conveys the use of information technology services and resources that are provided on a service basis.

译文：云计算是一个相对较新的术语，其含义是信息技术服务的使用和在服务基础上所提供资源的使用。

分析：句中"that conveys the use of information technology services and resources"是定语从句，修饰、限定先行词"term"；定语从句"that are provided on a service basis"修饰、限定先行词"resources"。

5. Cloud computing is a paradigm in which information is permanently stored in servers on the Internet and cached temporarily on clients that include desktops, entertainment centers, table computers, notebooks, wall computers, sensors, monitors, etc.

译文：云计算是一种模式，其中信息永久存储在互联网上的服务器中，临时信息缓存在客户端上，包括桌面、娱乐中心、台式计算机、笔记本电脑、壁挂式计算机、传感器、监视器等。

分析：句中"in which information is permanently stored in servers on the Internet and cached temporarily on clients"是定语从句，关系代词"which"指代"paradigm"；定语从句"that include desktops, entertainment centers, table computers, notebooks, wall computers, sensors, monitors, etc."修饰、限定先行词"clients"。

6. In the current (but still evolving) model of cloud computing, the cloud computing infrastructure consists of services that are offered up and delivered through data centers that can be accessed from anywhere in the world.

译文：在当前（但仍在发展中）的云计算模型中，云计算基础设施包括通过可以从世界任何地方访问的数据中心提供和交付的服务。

分析：句中短语"consists of"意思是"包括"；"that are offered up and delivered through data centers"是定语从句，修饰、限定先行词"services"；定语从句"that can be accessed from anywhere in the world"修饰、限定先行词"data centers"。

7. There will be more device and location independence, enabling users to access systems no matter where they are located or what kind of device they are using.

译文：将有更多的设备和位置独立性，使用户能够访问系统，无论他们在哪里或正在使用哪种设备。

分析：句中现在分词短语"enabling users to access systems"作状语，表示伴随；"no matter where...or what..."引导让步状语从句，意思是"无论哪里……或者无论什么……"。

Module 3   Consolidation Exercise 巩固练习

K6.PDF

I. Translate the following terms into English with the corresponding abbreviations

网络存储_____	集群文件系统_____
网络附加存储_____	分布式文件系统_____
云计算_____	存储区域网络_____
直连式存储_____	应用程序编程接口_____

II. Read the following statements and decide whether they are true or false according to the passage. Put "T" for True and "F" for False

1. These cloud storage users are responsible for keeping the data available and accessible, and the physical environment protected and running.

2. There are several approaches to clustering, all of which employ a clustered file system including Direct-Attached Storage for each node.

3. Typically, content-displaying hosting services (i.e., video and image), virtual storage, and remote backup allow HTTP access, and sometimes FTP access.

4. NAS is a file-level computer data storage server connected to a computer network offering data access to a heterogeneous group of clients.

5. Unlike SAN, NAS provides only block-based storage and leaves file system concerns on the "client" side.

6. Cloud computing is a paradigm in which information is temporarily stored in servers on the Internet and cached permanently on clients that include desktops, table computers, etc.

7. In the cloud computing definitions that are evolving, the services in the cloud are accessed by only enterprises and being provided by others via the Internet.

8. Efficiencies and cost savings around things like performance, load balancing, and even locations will be taken in to consideration in the sharing of costs and resources amongst so many users.

9. There is a big push for cloud computing services by several big companies, such as Amazon.com, Google and Microsoft.

III. Translate the following sentences into Chinese

1. People and organizations buy or lease storage capacity from the providers to store user, organization, or application data.

2. A clustered file system is a file system which is shared by being simultaneously mounted on multiple servers.

3. The term server highlights the role of the machine in the client–server scheme, where the clients are the workstations using the storage.

4. It is designed primarily to enable the storage and retrieval of data while the computation is carried out by the workstations.

5. However, file systems built on top of SANs do provide file-level access, and are known as shared-disk file systems.

6. In network diagrams, resources that are provided by an outside entity are depicted in a "cloud" formation.

7. Users of the services being offered often have very little knowledge of the technology being used.

8. Finally, cloud computing results in improved resource utilization, which is good for the sustainability movement (i.e., green technology or clean technology).

Unit Two   Network Engineering and Management
网络工程与管理

Lesson 7   Structured Cabling and Standardization Organization

Module 1   Text Study 课文学习

Basic Training 基本训练

Text 1   根据语音和视频完成以下任务。

Task 1-0 听录音，记录关键词，理解课文大意。

L7-1.MP3

L7-1.MP4

_____

_____

_____

Task 1-1 翻译、填空、词义辨析。

探索_____　研究_____　开发_____　设计_____

发布_____　部署_____　发展_____

manage 和 administrate 都是_____，但_____更常用一些，当同时出现时，_____的管理级别更高一些，Windows 系统的超级用户是_____，设备的管理员和密码常用_____作为默认值（default）。

Task 1-2 写出对应的句型。

按模块化方式设计。_____

以层次化方式集成。_____

在大多数情况下。_____

Task 1-3 uniform 和 unify 的动词语义分别是什么？

_____

Task 1-4 翻译。

Category 6 is rated for channel performance up to 200 MHz.

_____

Task 1-5 翻译以下专业相关的术语。

骨干网_____　单模光纤_____

多模光纤_____　屏蔽双绞线_____

非屏蔽双绞线_____　吉比特以太网_____

集成_____　额定性能_____

Task 1-6 写出英文全称。

① EIA_____

② TIA_____

Task 1-7 词义辨析。

feature_____　character_____

characteristic_____

课文

Structured Cabling

In the mid-1980s, the TIA (Telecommunications Industry Association) and the EIA (Electronic Industries Association) began developing methods for cabling buildings, with the intent of developing a uniform wiring system that would support multi-vendor products and environments. In 1991, the TIA/EIA released the TIA/EIA 568 Commercial Building Telecommunication Cabling Standard. The TIA/EIA structured cabling standards define how to design, build, and manage a cabling system that is structured, meaning that the system is designed in blocks that have very specific performance characteristics. The blocks are integrated in a hierarchical manner to create a unified communication system. For example, workgroup LANs represent a block with lower-performance requirements than the backbone network block, which requires high-performance fiber-optic cable in most cases. The standard defines the use of fiber-optic cable (single and multi-mode), STP (Shielded Twisted-Pair) cable, and UTP (Unshielded Twisted Pair) cable.

The current trend is to evolve the standards to support high-speed networking such as Gigabit Ethernet and define advanced cable types and connectors such as four-pair Category 6 and Category 7 cable. Category 6 is rated for channel performance up to 200 MHz, while Category 7 is rated up to 600 Mhz.

Text 2　根据语音和视频完成以下任务。

Task 2-0 听录音，记录关键词，理解课文大意。

_____

_____   L7-2-1.MP3

_____

_____   L7-2-2.MP3

_____

_____

_____   L7-2-3.MP3

_____

_____

_____   L7-2.MP4

Task 2-1 词义辨析。

① independent of 的近义词组是_____。

② limit 和 limitation 的区别是_____。

③ the last 和 the latest 的区别是_____。

Task 2-2 翻译画线部分。

① 插座_____由面板_____和底盒_____组成。

② 跳线_____，电信间_____，设备间_____，交联_____。

③ 50/125mm 是纤芯和光纤的直径，mm 表示"毫_____"，省略了米_____，即微米_____，一般写成 μm。

④ provider_____和 carrier_____在课文中都可翻译为_____。

⑤ 语音和数据_____。

Task 2-3 表示"指定"的词有_____和_____。

Task 2-4 翻译。

campus、building、entrance、floor、room、closet、wall box。

_____

Task 2-5 从 Text 1 和 Text 2 中找出以"multi-"为前缀和以"-tion"为后缀的词汇，理解词义。

_____

_____

Task 2-6 填空。

① The _____ length cannot _____ 10 meters (33 feet).

② There is no _____ on the number of telecommunication closets.

③ That UTP is _____ to 90 meters.

④ The backbone wiring provides the _____ for equipment rooms and telecommunication closets.

⑤ The distance _____ of this cabling depends on the type of cable.

Task 2-7 根据 Figure 2-1，用英语描述结构化布线系统的组成、结构、连接关系及距离特性。

_____

_____

_____

_____

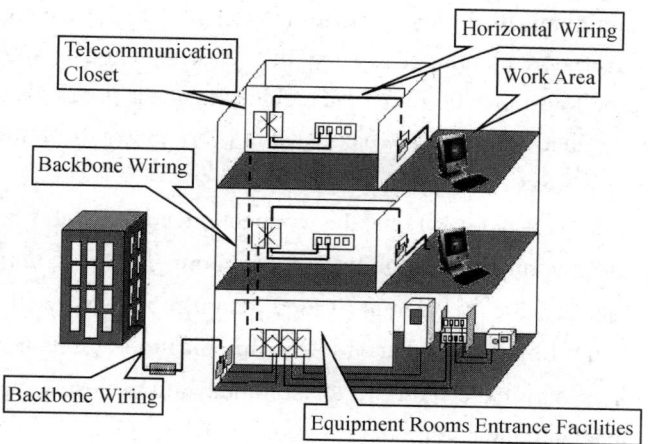

Figure 2-1   Structured Cabling

课文

The Definitions in TIA/EIA 568 Documents

According to TIA/EIA 568 documents, the wiring standard is designed to provide the following features and functions:

- A generic telecommunication wiring system for commercial buildings.
- Defined media, topology, termination and connection points, and administration.
- Support for multi-product, multi-vendor environments.
- Direction for future design of telecommunication products for commercial enterprises.
- The ability to plan and install the telecommunication wiring for a commercial building without any prior knowledge of the products that will use the wiring.

The layout of a TIA/EIA 568-A structured cable system is hierarchical. The hierarchical structure is apparent in the multi-floor office building. A vertical backbone cable runs from the central hub/switch in the main equipment room to a hub/switch in the telecommunication closet on each floor. Work areas are then individually cabled to the equipment in the telecommunication closet.

The TIA standard defines the parameters for each part of the cabling system, which includes work area wiring, horizontal wiring, telecommunication closets, equipment rooms and cross-connects, backbone (vertical) wiring, and entrance facilities.

Work Area Wiring Subsystem

The work area wiring subsystem consists of the communication outlets (wall boxes and faceplates), wiring, and connectors needed to connect the work area equipment (computers, printers, and so on) via the horizontal wiring subsystem to the telecommunication closet. The standard requires that two outlets be provided at each wall plate-one for voice and one for data.

Horizontal Wiring

The horizontal wiring system runs from each workstation outlet to the telecommunication closet. The maximum horizontal distance from the telecommunication closet to the communication outlets is 90 meters (295 feet) independent of media type. An additional 6 meters (20 feet) is allowed for patch cables at the telecommunication closet and at the workstation, but the combined length cannot exceed 10 meters (33 feet). As mentioned earlier, the work area must provide two outlets. The horizontal cable should be four-pair 100-ohm UTP cable (the latest standards specify Category 5e), or 50/125-mm multimode fiber-optic cable. Coaxial cable is no longer recommended.

Telecommunication Closet

The telecommunication closet contains the connection equipment for workstations in the immediate area and a cross-connection to an equipment room. The telecommunication closet is a general facility that can provide horizontal wiring connections, as well as entrance facility connections. There is no limit on the number of telecommunication closets allowed. Some floors in multistory office buildings may have multiple telecommunication closets, depending on the floor plan. These may be connected to an equipment room on the same floor.

Equipment Rooms and Main Cross-Connects

An equipment room provides a termination point for backbone cabling that is connected to one or more telecommunication closets. It may also be the main cross-connection point for the entire facility. In a campus environment, each building may have its own equipment room, to which telecommunication closet equipment is connected, and the equipment in this room may then be connected to a central campus facility that provides the main cross-connect for the entire campus.

Backbone Wiring

The backbone wiring runs up through the floors of the building or across a campus and provides the interconnection for equipment rooms and telecommunication closets. The distance limitations of this cabling depend on the type of cable and facilities it connects. Note that UTP is limited to 90 meters.

Entrance Facilities

The entrance facility contains the telecommunication service entrance to the building. This facility may also contain campus-wide backbone connections. It also contains the network demarcation point, which is the interconnection to the local exchange carrier's telecommunication facilities. The demarcation point is typically 12 inches from where the carrier's facilities enter the building, but the carrier may designate otherwise.

Advanced Training 进阶训练

T7.PDF

Task 1 阅读和理解下文，并翻译画线的词汇。

Standard Organizations Relating to Integrated Wiring

The International Organization for Standardization (ISO) is an 国际的＿＿＿＿＿＿＿ standard-setting body composed of representatives from various 国家的＿＿＿＿＿ standards organizations.

The International Electro technical Commission (IEC) is a 非营利性的＿＿＿＿＿, non-governmental international standards organization that prepares and publishes International Standards for all 电的＿＿＿＿＿, 电子的＿＿＿＿＿ and related technologies-collectively known as "electro technology".

The Institute of Electrical and Electronics Engineers (IEEE) is a professional association with its corporate office in New York City and its operations center in Piscataway, New Jersey. It is the world's largest association of technical professionals with more than 400,000 members in chapters around the world.

The American National Standards Institute (ANSI) is a private non-profit organization that oversees the development of voluntary consensus standards for products, services, processes, systems, and personnel in the United States.

The Telecommunications Industry Association (TIA) is accredited by the American National Standards Institute (ANSI) to develop voluntary, consensus-based industry standards for a wide variety of Information and Communication Technologies (ICT) products, and currently represents nearly 400 companies. TIA's Standards and Technology Department operates twelve engineering committees, which develop guidelines for private radio equipment, cellular towers, data terminals, satellites, telephone terminal equipment, VoIP devices, structured cabling, data centers, mobile device communications, multimedia 组播＿＿＿＿＿, vehicular telematics＿＿＿＿＿, healthcare ICT, machine to machine communications, and smart utility networks.

The Electronic Industries Alliance was a standards and trade organization composed as an alliance of trade associations for electronics manufacturers in the United States. They developed standards to ensure the equipment of different manufacturers was 兼容＿＿＿＿＿＿ and interchangeable＿＿＿＿＿＿.

CENELEC is responsible for European standardization in the area of electrical engineering. Together with ETSI (European Telecommunications Standards Institute) and CEN (other technical areas), it forms the European system for technical standardization.

The European Committee for Standardization is a public standards organization whose mission is to foster the economy of the European Union in global trading, the welfare of European citizens and the environment by providing an efficient 基础设施＿＿＿＿＿ to interested parties for the 开发＿＿＿＿＿, 维护＿＿＿＿＿ and 发布＿＿＿＿＿ of coherent sets of 标准＿＿＿＿＿ and

规范_____.

Task 2 找出表示组织、协会、联盟、委员会之类的词。

Organization、_____。

Vocabulary Practice 词汇练习

C7.MP3

词汇及短语听写，并纠正发音。

_____

_____

_____

_____

_____

Words & Expressions

multi-vendor [ˈmʌltɪˈvendə(r)]	*adj.* 多厂商（的）
generic [dʒəˈnerɪk]	*adj.* 一般的；类的，属性的
termination [ˌtɜːmɪˈneɪʃn]	*n.* 终止处
apparent [əˈpærənt]	*adj.* 可看见的；显然的；貌似的，表面的
specification [ˌspesɪfɪˈkeɪʃn]	*n.* 规格；详述；说明书
illustration [ˌɪləˈstreɪʃn]	*n.* 说明；例证；图解；插图
faceplate [ˈfeɪspleɪt]	*n.* 面板，花盘
exceed [ɪkˈsiːd]	*vt.* 超过，超越；胜过；越过……的界限
multistory [ˌmʌltɪˈstɔːrɪ]	*adj.* 多层的
demarcation [ˌdiːmɑːˈkeɪʃn]	*n.* 划界，立界
non-profit [nɒnˈprɒfɪt]	*adj.* 非营利性的
oversee [ˌəʊvəˈsiː]	*v.* 监督，监视
consensus [kənˈsensəs]	*n.* 一致
accredit [əˈkredɪt]	*v.* 相信；委托；委任
guideline [ˈgaɪdlaɪn]	*n.* 指导方针；指导原则
telematics [ˌtelɪˈmætɪks]	*n.* 信息技术
coherent [kəʊˈhɪərənt]	*adj.* 一致的

Module 2　Study Aid 辅助帮学

Terms & Abbreviations 术语与缩写

TIA　　*abbr.* 美国通信工业协会（Telecommunications Industry Association），是美国的一个全方位服务性国家贸易组织，其成员包括为世界各地提供通信和信息技术产品、系统和专业技术服务的 900 余家大小公司，协会成员有能力制造和供应现代通信网中应用的所有产品。

EIA　　*abbr.* 美国电子工业协会（Electronic Industries Association），美国电子行业标准制定者之一。EIA 创建于 1924 年，广泛代表了设计、生产电子元件、部件、通信系统和设备的制造商、工业界、政府和用户的利益，在提高美国制造商的竞争力方面起到了重要的作用。

backbone network　　骨干网，是用来连接多个区域或地区的高速网络。

telecommunication closet　　电信间，又称交接间、接线间或配线间，是放置电信设备、电缆、光缆、终端配线设备，并进行布线交接的一个专用场所。

communication outlets　　通信接线盒（也叫信息插座，即 information outlets），是水平布线与工作站之间的接口，例如，典型的 8 芯组合式插座。

wall box　　暗线箱、暗线盒、底盒。

wall plate　　墙壁插座、墙板插座。

patch cable　　跳接电缆，两端都有连接头的线缆，用于连接邻近的网络设备或电脑。

equipment room　　设备间，是在每幢建筑物的适当地点进行网络管理和信息交换的场地，主要安装建筑物配线设备、电话交换机、计算机主机设备，入口设施也可安放在一起。

cross-connect　　交接是指配线设备和信息通信设备之间采用接插软线或跳线上的连接器件相连的一种连接方式，是一种固定的连接方式。

integrated wiring　　集成布线，也称综合布线（generic cabling），是一种模块化的、灵活性极高的建筑物内或建筑群之间的信息传输物理通道。通过它可使语音设备、数据设备、交换设备及各种控制设备与信息管理系统连接起来，同时也使这些设备与外部通信网络相连。

ISO　　*abbr.* 国际标准化组织（International Organization for Standardization），是一个全球性的非政府组织，是国际标准化领域中一个十分重要的组织，负责目前绝大部分领域（包括军工、石油、船舶等垄断行业）的标准化活动。

IEC　　*abbr.* 国际电工委员会（International Electro technical Commission），是世界上成立最早的国际性电工标准化机构，负责有关电气工程和电子工程领域中的国际标准化工作。

IEEE　　*abbr.* 电气和电子工程师协会（Institute of Electrical and Electronics Engineers），是一个国际性的电子技术与信息科学工程师的协会，是目前全球最大的非营利性专业技术学会，其会员人数超过 40 万，遍布 160 多个国家。IEEE 致力于电气、电子、计算机工程和与科学有关领域的开发和研究，在太空、计算机、电信、生物医学、电力及消费性电子产品等领域已制定了 900 多个行业标准，现已发展成为具有较大影响力的国际学术组织。

ANSI　　*abbr.* 美国国家标准协会（American National Standards Institute）。

ICT　　*abbr.* 信息及通信技术（Information and Communication Technologies）。以前的通信技术与信息技术是两个相关性不高的范畴，通信技术着重于消息的传送技术，而信息技术着重于信息的编码、解码和在通信载体上的传输方式。随着技术的发展，这两种技术慢慢变得密不可分，渐渐融合为一个范畴。

VoIP　　*abbr.* 网络电话、互联网电话或 IP 电话（Voice over Internet Protocol），通过 IP 数据包发送来实现的语音业务，它将模拟的声音信号数字化、压缩与封包，以数据包的形式在 IP 网络中进行传输，通过互联网免费或资费很低地发送语音、传真、视频和数据等。

CENELEC　　欧洲电工标准化委员会（法文名称缩写），其宗旨是协调欧洲有关国家的标准机构所颁布的电工标准和消除贸易上的技术障碍。

CEN　　欧洲标准化委员会（法文名称缩写），是以西欧国家为主体、由国家标准化机构组成的非营利性标准化机构，其宗旨在于促进成员国之间的标准化协作，制定本地区需要的欧洲标准和协调文件。

Difficult Sentences Analysis and Translation 难句分析与翻译

1. In the mid-1980s, the TIA (Telecommunications Industry Association) and the EIA (Electronic Industries Association) began developing methods for cabling buildings, with the intent of developing a uniform wiring system that would support multi-vendor products and environments.

译文：在 20 世纪 80 年代中期，为了开发一个支持多厂商产品和环境的统一布线系统，TIA（美国通信工业协会）和 EIA（美国电子工业协会）开始研发建筑布线的方法。

分析：句中动名词短语"developing methods for cabling buildings"作动词宾语；介词短语"with the intent of"表示目的；定语从句"that would support multi-vendor products and environments"修饰、限定"a uniform wiring system"。

2. The TIA/EIA structured cabling standards define how to design, build, and manage a cabling system that is structured, meaning that the system is designed in blocks that have very specific performance characteristics.

译文：TIA / EIA 结构化布线标准定义了如何设计、构建和管理结构化的布线系统，这意味着系统应设计为具有非常明确的性能特点的模块。

分析：句中"how to design, build, and manage a cabling system"作动词宾语；定语从句"that is structured"修饰、限定"a cabling system"；现在分词短语"meaning that the system…"作伴随状语，表示同时发生；"that the system is designed in blocks"是宾语从句；定语从句"that have very specific performance characteristics"修饰、限定"blocks"。

3. In a campus environment, each building may have its own equipment room, to which telecommunication closet equipment is connected, and the equipment in this room may then be connected to a central campus facility that provides the main cross-connect for the entire campus.

译文：在校园环境中，每幢建筑物会有自己的设备间，电信间的设备连接到该设备间。该设备间里的设备连接到为整个校园提供主交叉连接的中心校园设施。

分析：句中"to which telecommunication closet equipment is connected"是非限制性定语从句，该从句中的关系代词"which"指代先行词"equipment room"；连词"and"连接前后两个并列的句子；定语从句"that provides the main cross-connect for the entire campus"修饰、限定"a central campus facility"。

4. TIA's Standards and Technology Department operates twelve engineering committees, which develop guidelines for private radio equipment, cellular towers, data terminals, satellites, telephone terminal equipment, VoIP devices, structured cabling, data centers, mobile device communications, multimedia multicast, vehicular telematics, healthcare ICT, machine to machine communications, and smart utility networks.

译文：TIA 的标准技术部有 12 个工程委员会在运作。这些工程委员会为私人无线电设备、蜂窝塔、数据终端、卫星、电话终端设备、VoIP 设备、结构化布线、数据中心、移动设备通信、多媒体组播、车载远程信息处理、医疗 ICT、机器对机器通信和智能多用途网络制定指南。

分析：本句术语较多，句式本身并不复杂。"which"一直引导非限制性定语从句到句末，关系代词"which"指代"twelve engineering committees"，翻译时单独成句。

5. The European Committee for Standardization is a public standards organization whose mission is to foster the economy of the European Union (EU) in global trading, the welfare of European citizens and the environment by providing an efficient infrastructure to interested parties for the development, maintenance and distribution of coherent sets of standards and specifications.

译文：欧洲标准化委员会是一个公共标准组织，其使命是通过为有关各方的研发、维护以及成套标准和规范的发布提供有效的基础设施，促进欧洲联盟（EU）在全球的经济贸易、欧洲公民的福利和环境。

分析：句中的定语从句"whose mission is to foster the economy of …"修饰、限定"a public standards organization"；介词短语"by providing an efficient infrastructure … and specifications"表示方式；短语"provide…for…"意思是"为……提供……"。

Module 3   Consolidation Exercise 巩固练习

I. Translate the following into English

K7.PDF

结构化布线_____   骨干网_____

电信间_____   工作区_____

通信接线盒_____   无线电设备_____

数据终端_____   卫星_____

电话终端设备_____   数据中心_____

移动设备通信_____   车载远程信息处理_____

交叉连接_____   网络分界点_____

标准组织_____   综合布线_____

II. Fill in each blank with appropriate words or expressions according to the passage

1. TIA/EIA structured cabling standards define how to design, build, and manage _____ _____ designed in blocks that are integrated in a _____ manner, thus creating a unified communication system.

2. The hierarchical structure is _____ in the multi-floor office building. A vertical backbone cable runs from the central hub/switch in the main equipment room to a hub/switch in _____ on each floor.

3. The standard requires that two _____ be provided at each _____one for voice and one for data.

4. The work area must provide two outlets. The horizontal cable should be _____ _____ (the latest standards specify Category 5e), two-fiber 62.5/125-mm fiber-optic cable, or _____ multimode fiber-optic cable.

5. The telecommunication closet contains the connection equipment for _____ in the immediate area and a _____ to an equipment room.

6. An equipment room _____ a termination point for backbone cabling that is _____ one or more telecommunication closets.

7. The entrance facility contains _____, which is the _____ to the local exchange carrier's telecommunication facilities.

8. The Electronic Industries Alliance developed standards to ensure the equipment of different manufacturers was _____ and _____.

III. Translate the following sentences into Chinese

1. The blocks are integrated in a hierarchical manner to create a unified communication system.

2. According to TIA/EIA 568 documents, the wiring standard is designed to provide the following features and functions.

3. The horizontal wiring system runs from each workstation outlet to the telecommunication closet.

4. An additional 6 meters (20 feet) is allowed for patch cables at the telecommunication closet and at the workstation.

5. Some floors in multistory office buildings may have multiple telecommunication closets, depending on the floor plan.

6. The distance limitations of this cabling depend on the type of cable and facilities it connects.

7. IEEE is the world's largest association of technical professionals with more than 400,000 members in chapters around the world.

8. The mission of the European Committee for Standardization is to foster the economy of the European Union in global trading, the welfare of European citizens and the environment.

Lesson 8   Cables, Connectors and Wireless Links

Module 1   Text Study 课文学习

Basic Training 基本训练

Text 1   根据语音和视频完成以下任务。

Task 1-0 听录音，记录关键词，理解课文大意。

_____

_____

_____   L8-1-1.MP3

_____

_____   L8-1-2.MP3

_____

Task 1-1 在 "Cat 5e has stretched the Cat 5 standard to its limits." 中 stretch 的   L8-1.MP4
近义词是_____。

Task 1-2 写出对应的中文或英文。

on the horizon_____   双绞线束_____   stand for_____

兆比特每秒_____或_____   屏蔽的_____或_____   绝缘的_____

天线_____   百兆以太网_____   千兆以太网_____

万兆以太网_____   电磁干扰_____

Task 1-3 词义辨析。

① property 和 performance_____

② use 和 apply_____

③ use 和 usage_____

Task 1-4 翻译。

① 这个板子有 1 毫米厚。_____

② 这个板子的厚度是 1 毫米。_____

③ 这根跳线有 1 米长。_____

④ 这根跳线的长度是 1 米。_____

⑤ 这个楼梯是螺旋式上升的。_____

⑥ 光纤的直径是 125 微米。_____

⑦ 这个校园网覆盖了数千台笔记本电脑。_____

⑧ 这种方法的应用使电磁干扰抵消了。_____

⑨ 之后的数年里，这个插座仍然是欧洲样式的。_____

⑩ 它最大运行速度为 1000 Mbps。_____

⑪它通过各种方法实现了 1000 兆的带宽。_____

Task 1-5 写出图（Figure 2-2）中各个部件的英文名称。

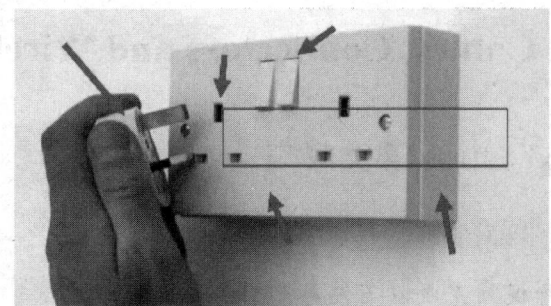

Figure 2-2　Analyzing Components

课文

Twisted Pair

One of the oldest and still most common transmission media is twisted pair. As shown in Figure 2-3, a twisted pair consists of two insulated copper wires, typically about 1mm thick. The wires are twisted together in a helical form. Twisting is done because two parallel wires constitute a fine antenna. When the wires are twisted, the waves from different twists cancel out, so the wire radiates less effectively.

Figure 2-3　UTP pairs (cable)

Twisted pairs can be used for transmitting either analog or digital signals. The bandwidth depends on the thickness of the wire and the distance traveled, but several Mb/s to tens of Mb/s can be achieved for a few kilometers in many cases. The frequency range of twisted-pair cables is approximately 0 to 10,000 MHz. Due to their adequate properties and low cost, twisted pairs are widely used and are likely to remain so for years to come.

Twisted pair cables are often shielded in attempt to prevent Electro-Magnetic Interference (EMI). Because the shielding is made of metal, it may also serve as a ground. However, usually a shielded or a screened twisted pair cable has a special grounding wire added called a drain wire. This shielding can be applied to individual pairs, or to the collection of pairs. When shielding is applied to the collection of pairs, this is referred to as screening. The shielding must be grounded for the shielding to work. In contrast to FTP (Foiled Twisted Pair) and STP (Shielded Twisted Pair) cabling, UTP (Unshielded Twisted Pair) cable is not surrounded by any shielding. It is the primary wire type for telephone usage and computer networking, especially as patch cables. UTP comes in several varieties:

Category 3: Was the earliest successful implementation of UTP. It's primarily used for voice and lower-speed data applications.

Category 5e: With the need for higher speeds, Gigabit Ethernet has become the new replacement for Fast Ethernet. To make it work, Cat 5e extends the life of Cat 5 cable. It can run at a maximum of 1,000 Mbps.

Category 6: Cat 5e can run at gigabit speeds, but with 10-Gigabit Ethernet on the horizon, Cat

5e has stretched the Cat 5 standard to its limits. Cat 6 can currently run at 1,000 Mbps (1Gbps). The Category 6 specification was released for publication very recently, however as designed. Category 6 cabling will be able to support speeds up to at least 10Gbps.

The standard connector for UTP cabling is an RJ-45 connector. This is a plastic connector that looks like a large telephone-style connector (see Figure 2-4). A slot allows the RJ-45 to be inserted only one way. RJ stands for Registered Jack.

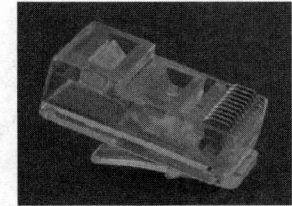

Figure 2-4   RJ-45 Connector

Text 2　根据语音和视频完成以下任务。

Task 2-0 听录音，记录关键词，理解课文大意。

L8-2-1.MP3

L8-2-2.MP3

L8-2-3.MP3

_____
_____
_____
_____
_____
_____
_____
_____

L8-2.MP4

Task 2-1 填空。
① 用来表示"组成的部分"的词有_____、_____、_____。
② "微米"的表达方式有_____、_____、_____、_____。

Task 2-2 翻译画线部分。
ultra 单独的意思是"过激分子、极端的（地）"，作为词素构成的词有：超细_____，超纯_____，紫外线_____，超声波_____等；类似的有由"infra-"前缀构成的红外线_____。

Task 2-3 写出对应的中文或英文。
生成_____　创造_____　形成_____
convert_____　change_____　turn to_____
损耗_____　失败者_____　串音_____
individual_____　separate_____

Task 2-4 翻译。
① 这个核心的半径是 5 微米。_____
② 光线沿光纤传播。_____
③ 一个脉冲代表比特 1，否则是比特 0。_____
④ 一个脉冲代表一个比特。_____

Task 2-5 词义辨析。
① safe 和 secure_____
② safety 和 security_____

Task 2-6 描述光缆的结构。

_____

_____

Task 2-7 用英文陈述单模光纤的优点和缺点。

_____

_____

_____

课文

Optical Fiber

An optical transmission system has three key components: the light source, the transmission medium, and the detector. Conventionally, a pulse of light indicates a 1 bit and the absence of light indicates a 0 bit. The transmission medium is an ultra-thin fiber of glass or plastic. The detector generates an electrical pulse when light falls on it. By attaching a light source to one end of an optical fiber and a detector to the other, we have a unidirectional data transmission system that accepts an electrical signal, converts and transmits it by light pulses, and then reconverts the output to an electrical signal at the receiving end. Higher bandwidth links can be achieved using optical fibers. One of the best substances used to make optical fibers is ultra-pure fused silica. These fibers are more expensive than regular glass fibers. Plastic fibers are normally used for short-distance links where higher losses are tolerable.

Optical fiber links are used in all types of networks, LAN and WAN. The frequency range of fiber optics is approximately 180 THz to 330 THz. There are two types of fiber optics cables: multi-mode fiber and single-mode fiber.

- Multi-mode fiber.

Light rays can only enter the core if their angle is inside the numerical aperture of the fiber. Once the rays have entered the core of the fiber, there are a limited number of optical paths that a light ray can follow through the fiber. These optical paths are called modes. If the diameter of the core of the fiber is large enough so that there are many paths that light can take through the fiber, the fiber is called "multi-mode" fiber. Single-mode fiber has a much smaller core that only allows light rays to travel along one mode inside the fiber.

Until the connectors are attached, there is no need for shielding, because no light escapes when it is inside a fiber. There are no crosstalk issues with fiber. It is very common to see multiple fiber pairs encased in the same cable. One cable can contain 2 to 48 or more separate fibers. Fiber can carry many more bits per second and carry them farther than UTP can.

Usually, five parts make up each fiber-optic cable. The parts are the core, the cladding, a buffer, a strength material, and an outer jacket.

Infrared Light Emitting Diodes (LEDs) types of light source is usually used with multi-mode fiber. LEDs are cheap to build and require somewhat less safety concerns than lasers. However, LEDs cannot transmit light over cable as far as the lasers. Multi-mode fiber (62.5/125) can carry data distances of up to 2 km.

- Single-mode fiber.

Single-mode fiber consists of the same parts as multi-mode. The outer jacket of single-mode fiber is usually yellow. The major difference between multi-mode and single-mode fiber is that single-mode allows only one mode of light to propagate through the smaller, fiber-optic core. The single-mode core is eight to ten μm in diameter. Nine-micron cores are the most common. A 9/125 marking on the jacket of the single-mode fiber indicates that the core fiber has a diameter of 9 microns and the surrounding cladding is 125 microns in diameter.

An infrared laser is used as the light source in single-mode fiber. The ray of light it generates enters the core at a 90-degree angle. The data carrying light ray pulses in single-mode fiber is essentially transmitted in a straight line right down the middle of the core. This greatly increases both the speed and the distance that data can be transmitted.

Single-mode fiber is capable of higher bandwidth and greater cable run distances than multi-mode fiber. Single-mode fiber can carry LAN data up to 3 km. Although this distance is considered a standard, newer technologies have increased this distance. Multi-mode is only capable of carrying up to 2 km. Lasers and single-mode fibers are more expensive than LEDs and multi-mode fibers. Because of these characteristics, single-mode fiber is often used for inter-building connectivity. Multi-mode and single-mode fibers are shown in Figure 2-5.

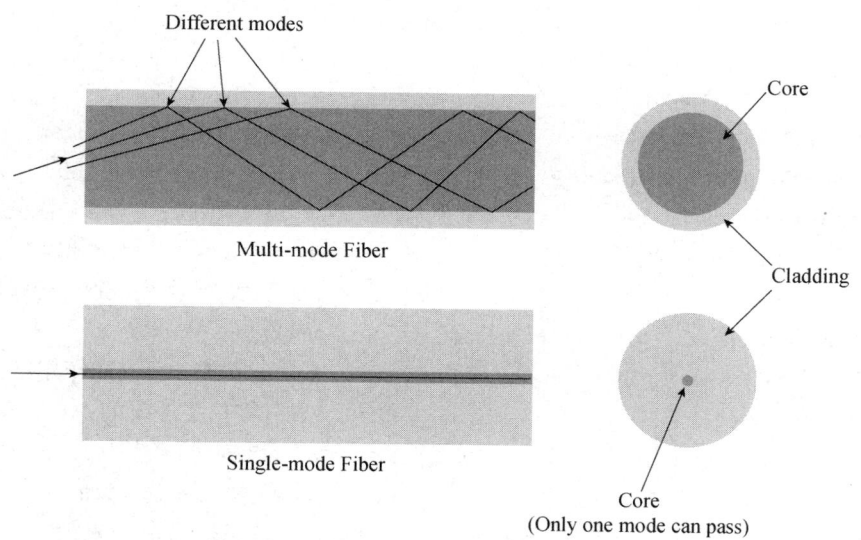

Figure 2-5   Multi-mode and Single-mode Fiber

Advanced Training 进阶训练

Wireless Communication

T8.PDF

Nowadays there are people who need to be on-line all the time. For these mobile users, twisted pair and fiber optics are of no use. They need to get data for their laptop or wristwatch computers without being tethered to the terrestrial communication infrastructure. For these users, wireless communication is the answer. Computer networks can take advantage of the wireless infrastructure where physical wires cannot be laid out. Some people believe that the future holds only two kinds of communications: fiber and wireless. All non-mobile computers, telephones, printers, and so on will use fiber, and all mobile ones will use wireless.

Task: Why is wireless communication necessary?

_____

_____

_____

Vocabulary Practice 词汇练习

词汇及短语听写，并纠正发音。

C8.MP3

_____

_____

_____

_____

Words & Expressions

insulated [ˈɪnsjuleɪtɪd]	*adj.* 绝缘的，隔热的
helical [ˈhelɪkl]	*adj.* 螺旋状的
antenna [ænˈtenə]	*n.* 天线；触角
radiate [ˈreɪdɪeɪt]	*vt.* 辐射；放射；流露
interference [ˌɪntəˈfɪərəns]	*n.* 干涉，干扰
conventionally [kənˈvenʃənəlɪ]	*adv.* 按照惯例，照常规
unidirectional [ˌjuːnɪdɪˈrekʃənəl]	*adj.* 单向的，单向性的
reconvert [ˈriːkənˈvɜːt]	*vt.* 重新转换；（使）恢复故态
aperture [ˈæpətʃə(r)]	*n.* 孔径；孔，洞；隙缝
encase [ɪnˈkeɪs]	*vt.* 包装；围绕；把……装箱
cladding [ˈklædɪŋ]	*n.* 覆层；保护层
buffer [ˈbʌfə(r)]	*n.* 缓冲器；起缓冲作用的人（或物）
infrared [ˌɪnfrəˈred]	*adj.* [物] 红外线的
diameter [daɪˈæmɪtə(r)]	*n.* 直径，直径长
micron [ˈmaɪkrɒn]	*n.* 微米（符号：μm）
tether [ˈteðə(r)]	*vt.* 系绳；拘束；束缚

Module 2　Study Aid 辅助帮学

Terms & Abbreviations 术语与缩写

EMI　　*abbr.* 电磁干扰（Electro-Magnetic Interference），是干扰信号并降低信号完好性的电子噪声，通常由电磁辐射发生源（如马达和机器）产生。

drain wire　　引流线，也叫加屏线，用于减少电磁干扰，从现场引到辅助柜的接地端，最后汇总到工作接地汇流排进入接地系统。

FTP　　*abbr*. 铝箔屏蔽双绞线（Foiled Twisted Pair），采用整体屏蔽结构，由多对双绞线外包裹铝箔构成。

Category 5e　　超五类，指超过五类双绞线的制造标准的类别。

Fast Ethernet　　快速以太网，数据速率可以达到100Mbps，是标准以太网（10Mbps）数据速率的十倍。

RJ　　*abbr*. 注册插孔（Registered Jack），国际标准8芯以太网线连接插头，来源于贝尔系统的通用服务分类代码（Universal Service Ordering Codes，USOC），是美国电子工业协会、美国通信工业协会确立的一种以太网连接器的接口标准。

optical fiber　　光纤，光导纤维的简写，是一种由玻璃或塑料制成的纤维，可作为光传导媒介，其传输原理是光的全反射。

fused silica　　熔融石英，是氧化硅（石英、硅石）的非晶态（玻璃态），是典型的玻璃。

multi-mode fiber　　多模光纤，能传输多种模式（波长）的光，光纤内光的路径也可以很多。

single-mode fiber　　单模光纤，只能传输一种模式（波长）的光，光纤内光的路径也是唯一的、沿光纤本身的路径。

numerical aperture　　数值孔径，光学系统的数值孔径是一个无量纲的数，用以衡量该系统能够收集的光的角度范围。

crosstalk　　串扰，指两条信号线之间因耦合而产生的相互电磁干扰。

LED　　*abbr*. 二极管（Light Emitting Diode），是一种能将电能转化为光能的半导体电子元件。

laser/LASER　　*abbr*. 激光（Light Amplification by Stimulated Emission of Radiation），指通过受激辐射而产生、放大的光，即受激辐射的光放大，特点是单色性极好，发散度极小，亮度（功率）可以达到很高。

Difficult Sentences Analysis and Translation 难句分析与翻译

1. Once the rays have entered the core of the fiber, there are a limited number of optical paths that a light ray can follow through the fiber.

译文：一旦光线进入了光纤的纤芯，光线可以穿过光纤的光路数量是有限的。

分析：句中"Once"的意思是"一旦"，引导条件状语从句；"that a light ray can follow"是定语从句，修饰、限定"optical paths"。

2. If the diameter of the core of the fiber is large enough so that there are many paths that light can take through the fiber, the fiber is called "multi-mode" fiber.

译文：如果光纤的纤芯直径足够大，光线可以穿过光纤的路径就会很多，则该光纤被称为"多模"光纤。

分析：句中"If the diameter of the core of the fiber is large enough... through the fiber"是条件状语从句；在"If"引导的条件状语从句中，"so that there are many paths"是结果状语从句；"that light can take through the fiber"是定语从句，修饰、限定"many paths"。

3. Until the connectors are attached, there is no need for shielding, because no light escapes when it is inside a fiber.

译文：在连接器连接之前不需要屏蔽，因为在光纤内部光不会泄漏。

分析：句中"Until"在否定句中的意思是"直至……才……"，引导时间状语从句；在由"because"引导的原因状语从句中，"when it is inside a fiber"是时间状语从句。

4. A 9/125 marking on the jacket of the single-mode fiber indicates that the core fiber has a diameter of 9 microns and the surrounding cladding is 125 microns in diameter.

译文：单模光纤护套上的 9/125 标记表示芯纤维的直径为 9 微米，周围的包层的直径为 125 微米。

分析：句中"that"引导宾语从句，作动词"indicates"的宾语；宾语从句"that the core fiber has a diameter of 9 microns and the surrounding cladding is 125 microns in diameter"中，连词"and"连接两个并列的句子。

Module 3　Consolidation Exercise 巩固练习

K8.pdf

I. Answer the following questions according to the texts

1. What does a twisted pair consist of ?

2. Why are the wires twisted together in twisted pairs?

3. What are the substances that are used to make optical fibers?

4. What is the frequency range of fiber optics?

5. Why is wireless communication necessary for us nowadays?

6. How far can single-mode fiber carry LAN data?

7. Why do some people believe that the future holds only two kinds of communications?

II. Translate the following terms into Chinese

patch cable_____　　multi-mode fiber_____

crosstalk_____　　EMI_____

bandwidth_____　　Fast Ethernet_____

detector_____　　laser_____

micron_____　　antenna_____

III. Writing practice

Write a summary of this lesson in about one hundred words.

Lesson 9　Network Devices

Module 1　Text Study 课文学习

Basic Training 基本训练

Text 1　根据语音和视频完成以下任务。

Task 1-0 听录音，记录关键词，理解课文大意。

_____

_____

_____

_____

_____

L9-1-1.MP3

L9-1-2.MP3

L9-1.MP4

Task 1-1 写出相应的英文或中文，了解相关专业背景。

芯片组_____　印制电路板_____　硬件编码_____　头部_____

尾部_____　全双工_____　半双工_____　单工_____

衰减_____　损耗_____　扭曲_____　存储转发_____

分割_____　数据包_____　数据帧_____　溢出_____

fragment-free_____　carrier wave signal_____

flow control_____　congestion control_____　流量_____　或_____

和 stop 对应的是_____，和 income 对应的是_____，而不是_____。

Task 1-2 翻译。

① Hubs operate at the Physical Layer, with a hint of the Physical Layer of the OSI reference model.

② 常见的网络设备有：调制解调器、网卡、中继器、集线器、收发器、网桥、交换机、路由器、防火墙和网关等。

③ This involves more complex routing decisions that are made in the context of the network as a whole yet not at the level of complexity that characterizes a router.

④ A repeater operates within the Physical Layer of the OSI reference model and regenerates analog or digital signals that are distorted by transmission loss due to attenuation.

Task 1-3 辨析词义。
① fragment/segment
② eliminate/reduce
③ architecture/structure/scheme
④ modulate/encode
⑤ trailer/tail/tailer
⑥ in the event/case/context
⑦ respectively/individually
Task 1-4 "MAC" 中的 "M" 和 "A" 分别指的是_____和_____。
Task 1-5 在 "a switch can exercise a flow control mechanism" 中，和 exercise 语义最接近的词是_____。
Task 1-6 用英文简要解释：交换机是如何减少拥塞的？

课文

Physical Layer and Data Link Layer Devices

In addition to the medium, networks may use devices such as modem, Network Interface Card (NIC), repeater, bridge, hub, switch, router, and gateway.

A modem (modulator-demodulator) is a network hardware device that modulates one or more carrier wave signals to encode digital information for transmission and demodulates signals to decode the transmitted information.

A Network Interface Card (NIC), also known as Network Interface Unit (NIU), is chipsets on printed circuit boards that provide physical access from the node to the LAN medium. The NIC is responsible for fragmenting the data transmission and formatting the data packets with the necessary header and trailer. A standard IEEE NIC contains a unique, hard-coded logical address (MAC address), which is included in the header of each data packet it transmits. NICs function at the Physical and Data Link Layers. Fast Ethernet NIC is shown as Figure 2-6.

Figure 2-6   Fast Ethernet NIC

A repeater is a network device used to regenerate a signal. It operates within the Physical Layer of the OSI reference model and regenerates analog or digital signals that are distorted by transmission loss due to attenuation. A repeater does not make intelligent decision concerning forwarding packets.

Hubs can be either active or passive. Passive hubs act simply as cable-connecting devices, while active hubs also serve as signal repeaters and are called "multi-port repeaters". Hubs operate at the Physical Layer, with a hint of the Physical Layer of the OSI reference model.

Bridges are relatively simple devices that connect LANs of the same architecture. Bridges operate at the lower two layers of the OSI reference model, providing Physical Layer and Data Link Layer connectivity. Bridges act as LAN repeaters where specified distance limitations are exceeded. Bridges have buffers so they can store and forward frames in the event that the destination link is congested with traffic.

Switches are network devices with basic frame store-and-forward capabilities that can support multiple simultaneous transmissions. Switches are also called "multi-port bridges" (Figure 2-7). Switches have the ability to read the target MAC addresses of the frames and forward them only to the appropriate port associated with the target device. Switches operate at the Physical and Data Link Layers of the OSI reference model Layers 1 and 2, respectively. Switches read the destination addresses of the frames, filtering and forwarding as appropriate, based on MAC addresses (Layer 2 address). Switches are fast and relatively inexpensive. Some switches make routing decisions based on IP addresses (Layer 3 address). Layer 3 switching involves a combination of switching and routing. This involves more complex routing decisions that are made in the context of the network as a whole yet not at the level of complexity that characterizes a router.

Figure 2-7   Switch

A switch does a great deal to reduce congestion in a number of ways. First, a switch can support multiple simultaneous transmissions. Second, switches serve to segment a network through filtering, as they forward traffic only to the port associated with the link to which the target device is connected. Thereby, that traffic does not contribute to congestion on other links or segments. Third, a switch can be equipped to buffer incoming frames until internal bus resources are available to process them. A switch can also be equipped to buffer outgoing frames until such time as the link to the next switch becomes available. Fourth, a switch can exercise a flow control mechanism, whereby it can advise a device to stop transmitting when its buffers are in danger of overflowing and then advise the device to resume transmission when the pressure on resources has been relieved. Fifth, store-and-forward and fragment-free switches variously eliminate or reduce the number of error frames. Finally, a switch supports full-duplex transmission, thereby eliminating data collisions associated with CSMA in an Ethernet environment, assuming that the station is directly connected to the switch rather than through a hub. This approach is the current best practice.

Text 2   根据语音和视频完成以下任务。

Task 2-0 听录音，记录关键词，理解课文大意。

L9-2-1.MP3

_____

_____

_____

L9-2-2.MP3

_____

_____

_____

Task 2-1 写出对应的英文或中文，注意辨析词义，归纳构词法。

L9-2.MP4

类似的_____ 不相似的_____ （完全）不同的_____

封装_____ 绝缘_____ encase_____

packet_____ 负载类型_____ 优先等级_____

最小代价（开销）_____ 最小路由延时_____ 子网_____

子集_____ 子系统_____ 底/顶三层_____

on the order of_____ 可编程路由策略_____

Task 2-2 填空，理解词义，学习专业知识。

① constant、continuous 和 continual 都可以翻译为"连续的"，但语义是有差别的。_____指"恒定不变"，"常数"也是这个词；_____指"多次重复"，虽然中断了，但还有后继：_____指"连续不断的"。

② _____和_____都是指"边界路由器、接入路由器"，位于一个自治区域的边界，例如，校园网，用于将局域网接入互联网。

③ in effect 和 in fact 都可翻译为"实际上"，但_____是指"达到了这种效果"，但黑客是有可能突破这种屏蔽效果的。

④ Policy-based routers can provide various levels of service based on _____ such as the _____ of the user, the terminal and the type of payload.

⑤ In effect, even the very existence of those resources is _____ from view.

⑥ Creation of such isolated subnets may serve _____ reasons of security or simply _____ a means of avoiding unnecessary congestion.

Task 2-3 翻译。

① Routers are multi-port devices with high-speed ports running at rates up to 155 Mbps or more and with high-speed internal buses that can be on the order of 1 Gbps in the aggregate.

_____

_____

② Users associated with a subnet may be afforded access to only a limited subset of network resources in the form of sites, links, hosts, files, databases, and applications.

_____

③ In addition to being limited in terms of access to such a resource, users of another subnet may be prevented from receiving data from it.

_____

课文

Router

Routers (Figure 2-8) are highly intelligent network devices that can support connectivity between both like and disparate LANs and can provide access to various WANs, such as Frame Relay, IP and ISDN.

Figure 2-8   Router

Routers typically operate at the bottom three layers of the OSI model using the Physical Layer, Data Link Layer, and Network Layer to provide connectivity, addressing, and switching. Routers also have the capability to operate at all seven layers of the OSI reference model, if so equipped (Figure 2-9).

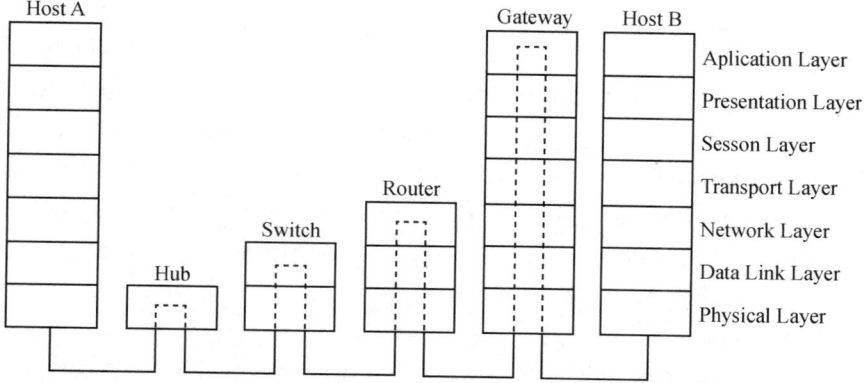

Figure 2-9   Networking Devices and OSI Model

In addition to supporting filtering and encapsulation, routers route traffic based on a high level of intelligence that enables them to consider the network as a whole. This is in stark contrast to bridges, hubs, and switches, which view the network simply on link-by-link basis. Routing considerations might include destination address, payload type, packet priority level, least-cost route, minimum route delay, minimum route distance, and route congestion level. Routers also are self-learning, as they can communicate their existence to other devices and can learn of the existence of new routers, nodes, and LAN segments. Routers constantly monitor the condition of the network as a whole; thereby dynamically adapting to changes in the condition of the network from edge to edge. Routers are multi-port devices with high-speed ports running at rates up to 155 Mbps or more and with high-speed internal buses that can be on the order of 1 Gbps in the aggregate. Additionally, routers typically provide some levels of redundancy so they are less susceptible to catastrophic failure.

Routers are unique in their ability to route data based on programmable network policy. Policy-based routers can provide various levels of service based on factors such as the identification of the user, the terminal and the type of payload. From one edge of the network to the other, an edge router can select the most appropriate path through the various switches or routers positioned in the core. An important part of this process is to divide the network into multiple subnets. Users associated with a subnet may be afforded access to only a limited subset of network resources in the form of sites, links, hosts, files, databases, and applications. In addition to being limited in terms of access to such a resource, users of another subnet may be prevented from receiving data from it. In effect, even the very existence of those resources is masked from view. Creation of such isolated subnets may serve for reasons of security or simply as a means of avoiding unnecessary congestion.

Advanced Training 进阶训练

T9.PDF

Task 1 将下列词汇以适当的形式填写在适当的位置
reside　　equivalent　　address　　division　　midrange　　functional　　mainframe
proprietary　　peak　　time-consuming　　perform

Gateways

Gateways can ＿＿＿＿＿＿＿（执行）all of the functions of switches and routers as well as accomplishing protocol conversion at all seven layers of the OSI reference model. Generally consisting of software ＿＿＿＿＿＿ in a host computer ＿＿＿＿＿＿ in processing power to a ＿＿＿＿＿（中型机）or ＿＿＿＿＿, gateway technology is expensive but highly ＿＿＿＿＿.

Protocol conversion, rather than encapsulation, can serve to fully convert from Ethernet to Token Ring or any other standard or ＿＿＿＿＿ protocol. Additionally, protocol conversion can ＿＿＿＿＿（处理）higher layers of the OSI model, perhaps through Layer 7, the Application Layer. As the process of protocol conversion is complex, gateways tend to operate rather slowly compared to switches and routers. As a result, they impose additional latency on packet traffic and may create bottlenecks of congestion during periods of ＿＿＿＿＿ usage. In a large and complex network, routers tend to be positioned at the edges of the network where they can be used to full advantage. Therefore, they make complex and ＿＿＿＿＿ decisions and invoke complex and time-consuming processes only where required. Switches tend to be positioned within the core of the network, because they can operate with greater speed.

Now we can conclude that switch creates separate collision domains and router creates separate broadcast domains. All ports of hub are one collision domain.

A collision domain is defined as a single CSMA/CD network segment in which there will be a collision if two computers attached to the system both transmit at the same time.

A broadcast domain is a logical _____ of a computer network, in which all nodes can reach each other by broadcast at the Data Link Layer. (Figure 2-10)

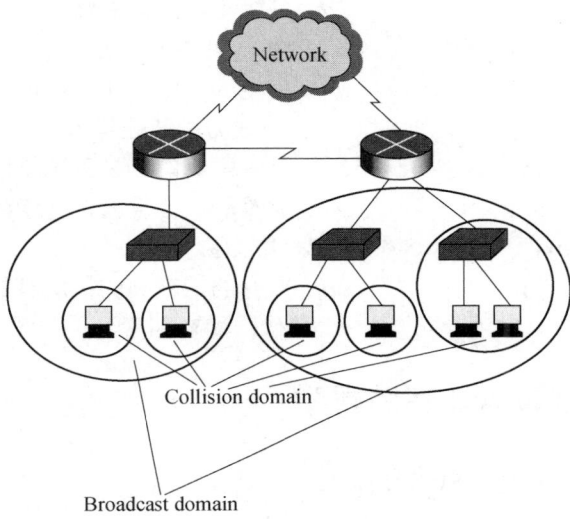

Figure 2-10　Collision Domain and Broadcast Domain

Task 2 翻译。

① In a large and complex network, routers tend to be positioned at the edges of the network where they can be used to full advantage.

_____

② Now we can conclude that switch creates separate collision domains and router creates separate broadcast domains. All ports of hub are one collision domain.

_____

Vocabulary Practice 词汇练习

词汇及短语听写，并纠正发音。

C9.MP3

_____
_____
_____
_____

Words & Expressions

fragment [ˈfrægmənt]　　　　　*vt.*（把）分段；（使）碎裂，分裂

attenuation [əˌtenjʊˈeɪʃn]　　　*n.* 变薄；弄细；稀薄化；减少

congest [kənˈdʒest]	*vt. & vi.* 拥挤；堵塞
appropriate [əˈprəʊprɪət]	*adj.* 适当的；恰当的；合适的
segment [ˈsegmənt]	*vt. & vi.* 分割，划分　*n.* [计] 段落；分段；部分
congestion [kənˈdʒestʃən]	*n.* 阻塞；充血；拥挤，堵车
mechanism [ˈmekənɪzəm]	*n.* 机制，机能
relieve [rɪˈliːv]	*vt.* 解除；缓解
eliminate [ɪˈlɪmɪneɪt]	*vt.* 排除，消除
disparate [ˈdɪspərət]	*adj.* 完全不同的；迥然不同的；无法比较的
encapsulation [ɪnˈkæpsjuleɪʃən]	*n.* 包装，封装，包裹
stark [stɑːk]	*adj.* 完全的；明显的　*adv.* 完全地；明显地
payload [ˈpeɪləʊd]	*n.* 有效载荷；装载量
susceptible [səˈseptəbl]	*adj.* 易受影响的；易受感染的
catastrophic [ˌkætəˈstrɒfɪk]	*adj.* 灾难的；惨重的，悲惨结局的
midrange [ˈmɪdˌreɪndʒ]	*n.* 适中范围
proprietary [prəˈpraɪətrɪ]	*adj.* 专有的，专利的；所有的；所有权的
latency [ˈleɪtənsɪ]	*n.* 潜伏；潜在因素

Module 2　Study Aid 辅助帮学

Terms & Abbreviations 术语和缩写

modem　调制解调器，是调制器（modulator）与解调器（demodulator）的统称。根据 Modem 的谐音，亲昵地称为"猫"。它是在发送端通过调制将数字信号转换为模拟信号，而在接收端通过解调，将模拟信号转换为数字信号的一种装置。

MAC　*abbr.* 媒体访问控制（Media Access Control）（子层协议），该协议位于 OSI 七层协议中数据链路层的下半部分。该协议定义了 48 位全球唯一的地址，按厂商分类，固化在网卡串行 EEPROM 上，非专用设备不能修改，所以也称为物理地址、硬件地址。

buffer　缓冲、缓存，指临时存储数据的存储器，可以起到调节、平衡输入与输出数据流量的作用。Cache 也是缓存，但一般指高速缓存，用于存储低速存储器的数据副本。

frame　帧，数据链路层的数据封装单位。

store and forward　存储和转发，先完整地接收数据于缓存中，以便对数据进行必要解读（如地址）和处理（如数据的完整性），再转发出去。

full-duplex　全双工，指信道可以同时接收和发送数据。

policy-based router　策略路由，一种比基于目标网络进行路由更加灵活的数据包的路由和转发机制。

Frame Relay　帧中继，一种面向分组的通信方法，主要用在公共或专用网上的网络互联技术。帧中继提供的是数据链路层和物理层的协议规范，任何高层协议都独立于帧中继协议。

Difficult Sentences Analysis and Translation 难句分析与翻译

1. A modem (modulator-demodulator) is a network hardware device that modulates one or more

carrier wave signals to encode digital information for transmission and demodulates signals to decode the transmitted information.

译文：调制解调器（调制器—解调器）是一种网络硬件设备。该设备调制一个或多个载波信号来对数字信息进行编码以便传输，并解调接收的信号来解码所传输的信息。

分析：句中定语从句"that modulates one or more carrier wave signals… to decode the transmitted information"修饰、限定"a network hardware"；不定式短语"to encode digital information"和"to decode the transmitted information"表示目的，作目的状语。

2. Bridges have buffers so they can store and forward frames in the event that the destination link is congested with traffic.

译文：网桥有缓存，因此，一旦目标链路拥塞，缓存就可以存储和转发帧。

分析：句中"so they can store and forward frames…"是目的状语从句；"in the event that the destination link is congested with traffic"是条件状语从句。

3. Fourth, a switch can exercise a flow control mechanism, whereby it can advise a device to stop transmitting when its buffers are in danger of overflowing and then advise the device to resume transmission when the pressure on resources has been relieved.

译文：第四，交换机可以执行流控机制。因此，当缓冲器有溢出危险时，它能建议设备停止发送；当资源上的压力已缓解时，则建议设备恢复传输。

分析：句中关系副词"whereby"相当于"as a result of which"，意思是"由此，因此，凭此"。

4. Finally, a switch supports full-duplex transmission, thereby eliminating data collisions associated with CSMA in an Ethernet environment, assuming that the station is directly connected to the switch rather than through a hub.

译文：最后，交换机支持全双工传输，因此如果该站不是通过集线器而是直接连接到交换机的话，可以在以太网环境中消除与 CSMA 相关的数据冲突。

分析：句中"thereby"的意思是"由此，因此"；短语"eliminating data collisions…"是用作结果状语的分词短语；"assuming that"是用作条件状语的分词短语，相当于 if 的用法，因此"assuming that the station is directly connected…"是条件状语从句。

Module 3   Consolidation Exercise 巩固练习

K9.PDF

I. Answer the following questions according to the texts

1. Which devices may networks use?

2. What is a modem?

3. What is the NIC responsible for?

4. Does a repeater make intelligent decisions concerning forwarding packets?

5. What are switches also called?

6. Why do switches serve to segment a network through filtering?

7. What might routing considerations include?

8. Can gateways perform all of the functions of switches and routers?

II. Translate the following sentences into Chinese

1. A Network Interface Card (NIC), also known as Network Interface Unit (NIU) is a chipset on printed circuit boards that provides physical access from the node to the LAN medium.

2. A standard IEEE NIC contains a unique, hard-coded logical address (MAC address), which is included in the header of each data packet it transmits.

3. Passive hubs act simply as cable-connecting devices, while active hubs also serve as signal repeaters and are called "multi-port repeaters".

4. Switches are network devices with basic frame store-and-forward capabilities that can support multiple simultaneous transmissions.

5. Routers typically operate at the bottom three layers of the OSI model using the Physical Layer, Data Link Layer, and Network Layer to provide connectivity, addressing, and switching.

6. Protocol conversion, rather than encapsulation, can serve to fully convert from Ethernet to Token Ring or any other standard or proprietary protocol.

7. A collision domain is defined as a single CSMA/CD network segment in which there will be a collision if two computers attached to the system both transmit at the same time.

III. Fill in each blank with appropriate words or expressions according to the passage

1. The NIC is responsible for _____ the data transmission and formatting the data packets with the necessary _____ and _____.

2. It operates within the Physical Layer of the OSI reference model and _____ analog or digital signals that _____ by transmission loss due to attenuation.

3. Bridges are relatively simple devices that connect LANs of the same _____.

4. In addition to _____ filtering and encapsulation, routers route traffic based on a high level of _____ that enables them _____ the network as a whole.

5. Routers are_____with high-speed ports running at rates up to 155 Mbps or more and with high-speed internal _____ that can be on the order of 1 Gbps in the aggregate.

6. Policy-based routers can provide various levels of service based on factors such as the _____ of the user, the terminal and the type of _____.

7. As the process of protocol conversion is complex, _____ tend to operate rather slowly compared to _____ and _____.

Lesson 10   Network Security

Module 1   Text Study 课文学习

Basic Training 基本训练

Text 1   根据语音和视频完成以下任务。

Task 1-0 听录音，记录关键词，理解课文大意。

_____   L10-1-1.MP3

_____

_____

_____   L10-1-2.MP3

_____

_____

Task 1-1 翻译画线的部分。   L10-1.MP4

网络安全_____有四个方面：鉴别_____、保密_____，完整性控制_____、反拒认_____。

加密_____   解密_____   窃听_____   伪造_____   更改_____

签名_____   弱点_____   敏感的_____   私密的_____   触发一个警报_____

Task 1-2 填空。

① "保密"和"安全"都有共同的词根_____，表示"分离"，而"-cy"是常见的名词结尾。

② deal with（处理）可用_____或 be _____替代。

③ principle 指_____，它的复数是_____，翻译时可以根据中文习惯酌情选择。但短语"原则上（in principle、by principle）"用单数。"原理"还可以用_____，其逻辑性更强。

Task 1-3 翻译。

① This is what usually comes to mind when people think about network security.

_____

② IP security also functions in this layer.

_____

③ Finally, integrity control has to do with how you can be sure that a message you received was really the one sent and not something that a malicious adversary modified in transit or concocted.

_____

④ How do you prove that your customer really placed an electronic order for ten million gloves at 90 cents each when he later claimed the price was 60 cents?

_____

Task 1-4 写出 6 个和"crypto-"相关的词（中、英文）。

_____

课文

Secrecy, Authentication, Nonrepudiation, and Integrity Control

Now, as millions of ordinary citizens are using networks for banking, shopping, and filing their tax returns, weakness after weakness has been found, and network security has become a problem of massive proportions. Network security problems can be divided roughly into four closely intertwined areas: secrecy, authentication, nonrepudiation, and integrity control. Secrecy, also called confidentiality, has to do with keeping information out of the grubby little hands of unauthorized users. This is what usually comes to mind when people think about network security. Authentication deals with determining whom you are talking to before revealing sensitive information or entering into a business deal. Nonrepudiation deals with signatures. How do you prove that your customer really placed an electronic order for ten million gloves at 90 cents each when he later claimed the price was 60 cents? Or maybe he claimed he never placed any order. Finally, integrity control has to do with how you can be sure that a message you received was really the one sent and not something that a malicious adversary modified in transit or concocted.

Except for Physical Layer security, nearly all network security is based on cryptographic principles. In the Physical Layer, wiretapping can be foiled by enclosing transmission lines (or better yet, optical fibers) in sealed tubes containing an inert gas at high pressure. Any attempt to drill into a tube will release some gas, reducing the pressure and triggering an alarm. In the Data Link Layer, packets on a point-to-point line can be encrypted as they leave one machine and decrypted as they enter another. In the network layer, firewalls can be installed to keep good packets in and bad packets out. IP security also functions in this layer. In the transport layer, entire connections can be encrypted end to end, that is, process to process. For maximum security, end-to-end security is required. Finally, issues such as user authentication and nonrepudiation can only be handled in the application layer.

Text 2 根据语音和视频完成以下任务。

Task 2-0 听录音，记录关键词，理解课文大意。

L10-2-1.MP3

_____

_____

_____

L10-2-2.MP3

_____

_____

_____

Task 2-1 填空。

L10-2.MP4

cryptography（密码学）和 encryption（加密术）都有词根_____。cipher 作为名词的意思是"密码"和"加密"，动词是"加密"，ciphers 指_____，词源是_____的意思，表示一种神秘的事物，引申为"计算、密码"。

Task 2-2 写出对应的英文词汇。

逆向_____　　对立面_____　　明文_____

密文_____　　字符串_____　　智能_____

智能的_____　　可理解的_____　　加密系统_____

形式上_____　　在实际操作中_____

Task 2-3 翻译。

① Formally, a "cryptosystem" is the ordered list of elements of finite possible plaintexts, finite possible cypher texts, finite possible keys, and the encryption and decryption algorithms which correspond to each key.

_____

_____

② Keys are important both formally and in actual practice, as ciphers without variable keys can be trivially broken with only the knowledge of the cipher used and are therefore useless (or even counter-productive) for most purposes.

_____

_____

Task 2-4 process、procedure 和 program 的词性和词义分别有什么不同？

_____

Task 2-5 用英语解释对称和非对称加密系统的优缺点。

_____

_____

_____

课文

Cryptography

Cryptography refers almost exclusively to encryption, which is the process of converting ordinary information (called plaintext) into unintelligible text (called cipher text). Decryption is the reverse, in other words, moving from the unintelligible cipher text back to plaintext. A cipher (or cypher) is a pair of algorithms that create the encryption and the reversing decryption. The detailed operation of a cipher is controlled both by the algorithm and in each instance by a "key". The key is a secret (ideally known only to the communicants), usually a short string of characters, which is needed to decrypt the cipher text. Formally, a "cryptosystem" is the ordered list of elements of finite possible plaintexts, finite possible cypher texts, finite possible keys, and the encryption and decryption algorithms which correspond to each key. Keys are important both formally and in actual practice, as ciphers without variable keys can be trivially broken with only the knowledge of the cipher used and are therefore useless (or even counter-productive) for most purposes. Historically, ciphers were often used directly for encryption or decryption without additional procedures such as authentication or integrity checks.

There are two kinds of cryptosystems: symmetric and asymmetric. In symmetric systems the same key (the secret key) is used to encrypt and decrypt a message. Data manipulation in symmetric systems is faster than asymmetric systems as they generally use shorter key lengths. Asymmetric systems use a public key to encrypt a message and a private key to decrypt it. Use of asymmetric systems enhances the security of communication. Examples of asymmetric systems include RSA (Rivest-Shamir-Adleman), and ECC (Elliptic Curve Cryptography). Symmetric models include the commonly used AES (Advanced Encryption System) which replaced the older DES (Data Encryption Standard).

Advanced Training 进阶训练

T10.PDF

Task 1 将下列词汇以适当形式填入画线处。

granular　implement　port number　combination　entire　categorize　exercise control　permit　deny　evaluate　field　entering or leaving　criteria　specified　reject　apply　examine　check　block　drawback　manipulate　signature　address　proxy

ACL

An Access Control List is essentially a list of conditions _____（分类）that packets. They can be really helpful when you need to _____（执行控制）over network traffic. One of the most common and easiest to understand the uses of access lists is filtering unwanted packets when _____（实施）security policies. There are two main types of Access Control Lists.

Standard access lists. These use only the source IP address in an IP packet as the condition test. All decisions are made based on the source IP address. This means that standard access lists basically _____（允许）or _____（拒绝）an _____（全部）suite of protocols. They don't distinguish between any of the many types of IP traffic such as Web, Telnet, UDP, and so on.

Extended access lists. Extended access lists can evaluate many of the other fields in the layer 3 and layer 4 headers of an IP packet. They can _____（评估）source and destination IP addresses, the protocol _____（区域）in the Network layer header, and the _____（端口号）at the Transport layer header. This gives extended access lists the ability to make many more _____（细分的）decisions when controlling traffic.

Firewall

A firewall is a system designed to prevent unauthorized access to or from a private network. You can implement a firewall in either hardware or software form, or a _____（结合）of both. Firewalls prevent unauthorized Internet users from accessing private networks connected to the Internet, especially Intranets. All messages _____（进入或离开）the Intranet (i.e., the local network to which you are connected) must pass through the firewall, which _____（审查）each message and _____（阻止）those that do not meet the _____（指定的）security _____（条件）.

Several types of firewall techniques exist:

Packet filtering. The system examines each packet entering or leaving the network and accepts or _____（抛弃）it based on user-defined rules. Packet filtering is fairly effective and transparent to users, but it is difficult to configure. In addition, it is susceptible to IP spoofing.

Circuit-level gateway implementation. This process _____（应用）security mechanisms when a TCP or UDP connection is established. Once the connection has been made, packets can flow between the hosts without further _____（检查）.

Acting as a _____（代理）server. A proxy server is a type of gateway that hides the true network address of the computer(s) connecting through it. A proxy server connects to the Internet, makes the requests for pages, connections to servers, etc., and receives the data on behalf of the computer(s) behind it. The firewall capabilities lie in the fact that a proxy can be configured to allow only certain types of traffic (e.g., HTTP files, or Web pages) to pass through. A proxy server has the potential _____（缺点）of slowing network performance, since it has to actively analyze and _____（操控）traffic passing through it.

IDS

Intrusion Detection Systems (IDSs) were implemented to passively monitor the traffic on a network. An IDS-enabled device copies the traffic stream, and analyzes the monitored traffic rather than the actual forwarded packets. It compares the captured traffic stream with known malicious _____（鲜明特征）in an offline manner similar to software that checks for viruses. This offline IDS implementation is referred to as promiscuous mode.

IPS

An Intrusion Prevention System (IPS) builds upon IDS technology. Unlike IDS, an IPS device is implemented in inline mode. This means that all ingress and egress traffic must flow through it for processing. An IPS does not allow packets to enter the trusted side of the network without first being analyzed. It can detect and immediately _____（定位并解决）a network problem as required.

The biggest difference between IDS and IPS is that an IPS responds immediately and does not allow any malicious traffic to pass, whereas IDS might allow malicious traffic to pass before responding. IDS and IPS technologies do share several characteristics. IDS and IPS technologies are both deployed as sensors.

Task 2 翻译。

① A firewall is a system designed to prevent unauthorized access to or from a private network.

_____

② Packet filtering is fairly effective and transparent to users, but it is difficult to configure. In addition, it is susceptible to IP spoofing.

_____

③ This offline IDS implementation is referred to as promiscuous mode.

_____

④ This means that all ingress and egress traffic must flow through it for processing.

_____

⑤ IDS and IPS technologies do share several characteristics. IDS and IPS technologies are both deployed as sensors.

_____

Vocabulary Practice 词汇练习

词汇及短语听写，并纠正发音。

C10.MP3

_____
_____
_____
_____
_____

Words & Expressions

proportion [prəˈpɔ:ʃn]	_n._ 比率；面积；份额	
nonrepudiation [ˈnʌnˈrepjuːdɪeɪʃən]	_n._ 不可否认，反拒认；认可	
confidentiality [ˌkɒnfiˌdenʃiˈæləti]	_n._ 机密性；保密性	
grubby [ˈɡrʌbi]	_adj._ 污秽的，邋遢的；肮脏的，不洁的	
reveal [rɪˈviːl]	_vt._ 显露；揭露；泄露	
unauthorized [ʌnˈɔːθəraɪzd]	_adj._ 未经授权的；未经许可的；未经批准的	
malicious [məˈlɪʃəs]	_adj._ 恶意的，有敌意的；蓄谋的；存心不良的	
adversary [ˈædvəsəri]	_n._ 对手，敌手	
concoct [kənˈkɒkt]	_vt._ 编造；捏造；调制	
cryptographic [ˈkrɪptəʊˈɡræfɪk]	_adj._ 关于暗号的，用密码写的	
wiretap [ˈwaɪətæp]	_vt. & vi._ 搭线窃听，窃听或偷录	
inert [ɪˈnɜːt]	_adj._ 迟钝的；不活泼的；惰性的	
encrypt [ɪnˈkrɪpt]	_vt._ 加密；把……加密（或编码）	
foil [fɔɪl]	_n._ 箔，金属薄片；用…陪衬，衬托；铺箔于	
granular [ˈɡrænjələ(r)]	_adj._ 颗粒状的；粒状的	
criteria [kraɪˈtɪəriə]	_n._（批评、判断等的）标准，准则（criterion 的复数）	
transparent [trænsˈpærənt]	_adj._ 透明的；清澈的（注：在 IT 领域，指从模	

块或层次外部来看，内部的技术细节不可见）

spoof [spu:f]	*n. & vt.* 滑稽的模仿；哄骗；戏弄
proxy ['prɒksɪ]	*n.* 代理服务器；代表权；代理人；委托书
drawback ['drɔ:bæk]	*n.* 缺点，劣势
manipulate [məˈnɪpjuleɪt]	*vt.* 操作，处理；巧妙地控制；操纵
promiscuous [prəˈmɪskjuəs]	*adj.* 混杂的；随便的
ingress ['ɪngres]	*n.* 进入；进入权；入口
egress ['i:gres]	*n.* 外出；出路，出口；运出
deploy [dɪˈplɔɪ]	*vt. & vi.* 有效地利用；（尤指军事行动）使施展
cryptography [krɪpˈtɒgrəfɪ]	*n.* 密码使用法，密码系统，密码术；密码学
exclusively [ɪkˈsklu:sɪvlɪ]	*adv.* 唯一地；专门地，特定地
decryption [di:ˈkrɪpʃn]	*n.* 解密，译码，密电码回译
unintelligible [ˌʌnɪnˈtelɪdʒəbl]	*adj.* 不能理解的，莫名其妙的
ciphertext ['saɪfətekst]	*n.* 密码，暗记文；密文
cipher ['saɪfə(r)]	*n.* 密码 *vt.* 用密码书写
algorithm ['ælgərɪðəm]	*n.* 运算法则；演算法；计算程序
communicant [kəˈmju:nɪkənt]	*n.* 通知者 *adj.* 传达的；通信的
finite ['faɪnaɪt]	*adj.* 有限的 *n.* 有限性；有限的事物
formal ['fɔ:ml]	*adj.* 正规的；正式的；庄重的
trivial ['trɪvɪəl]	*adj.* 琐碎的，无价值的；平凡的；不重要的
counter-productive [ˌkaʊntə prəˈdʌktɪv]	*adj.* 产生相反效果的；事与愿违的；适得其反的
symmetric [sɪˈmetrɪk]	*adj.* 对称的
asymmetric [ˌeɪsɪˈmetrɪk]	*adj.* 不对称的，不匀称的

Module 2   Study Aid 辅助帮学

Terms & Abbreviations 术语和缩写

integrity control   完整性控制，即确保数据有效、可用，没有被破坏和篡改。

ACL   *abbr.* 访问控制列表（Access Control List），是路由器用来控制数据包在端口能否进出的一个规则列表，根据 IP 地址和协议类型来决定是允许的或禁止的。

IDS   *abbr.* 入侵检测系统（Intrusion Detection System），依照一定的安全策略，通过软、硬件，对网络、系统的运行状况进行监测，尽可能发现各种攻击企图、攻击行为或者攻击结果，以保证网络系统资源的安全。

IPS   *abbr.* 入侵防御系统（Intrusion Prevention System），是对防病毒软件（Antivirus Programs）和防火墙（Firewall）的补充，用于监测网络中一些不正常或具有伤害性的网络行为。

RSA   *abbr.* 公钥加密算法（Rivest-Shamir-Adleman），是目前最有影响力的加密算法，能够抵抗目前已知的绝大多数密码攻击，已被 ISO 推荐为公钥数据加密标准。

ECC   *abbr.* 椭圆加密算法（Elliptic Curve Cryptography），是一种公钥加密机制，其数学基础是利用椭圆曲线上的有理点构成 Abel 加法群上椭圆离散对数的计算困难性。

AES　　*abbr.* 高级加密标准（Advanced Encryption Standard），在密码学中又称 Rijndael 加密法，是美国联邦政府采用的一种区块加密标准。这个标准用来替代原先的 DES，已经被多方分析且为全世界所使用。

DES　　*abbr.* 数据加密标准（Data Encryption Standard），是一种使用密钥加密的块算法，被美国联邦政府的国家标准局确定为联邦资料处理标准（FIPS），并授权在非密级政府通信中使用，随后该算法在国际上广泛流传开来。

Difficult Sentences Analysis and Translation 难句分析与翻译

1. Now, as millions of ordinary citizens are using networks for banking, shopping, and filing their tax returns, weakness after weakness has been found, and network security has become a problem of massive proportions.

译文：现在，由于数以百万计的普通公民都在使用网络开展银行业务、购物和税务申报，缺陷一个接一个被发现，而网络安全成为其中占比很大的一个问题。

分析：句中"as millions of ordinary citizens are using networks for banking, shopping, and filing their tax returns"是原因状语，主句是由 and 连接的两个并列分句。

2. Cryptography refers almost exclusively to encryption, which is the process of converting ordinary information (called plaintext) into unintelligible text (called ciphertext).

译文：密码学几乎只涉及加密，就是将普通信息（也称明文）转换成难以理解的文本（也称密文）的过程。

分析：句中"which"引导非限制性定语从句，对主句进行说明和解释。

3. The key is a secret (ideally known only to the communicants), usually a short string of characters, which is needed to decrypt the ciphertext.

译文：密钥是一个机密（理想情况下只有传递信息的人知道），通常是一个短的字符串，解密密文时需要密钥。

分析：句中插入语"usually a short string of characters"是同位语，补充说明前面的名词"secret"；"which"引导非限制性定语从句，对先行词进行附加说明和解释。

Module 3　Consolidation Exercise 巩固练习

K10.PDF

I. Translate the following sentences into English

1. 网络安全问题大致上分为四个紧密相连的领域：保密、鉴别、反拒认和完整性控制。

2. 身份验证用于在透露敏感信息或进行交易之前确定你在和谁交谈。

3. 在传输层，可以对全部的链接进行端到端加密。

4. 和入侵检测系统不同，入侵防御系统设备以串联模式部署。

5. 代理服务器有潜在降低网络性能的缺陷。

6. 加密方法有两种：对称和非对称。

7. 对称模型包括常用的高级加密系统，该系统取代了旧的数据加密标准。

II. Give the meaning of the following abbreviated terms both in English and in Chinese

ACL　　　　　　IDS　　　　　　DES　　　　　　AES
ECC　　　　　　RSA　　　　　　IPS

III. Talk about the ways of ensuring the security of network

Lesson 11　VLAN and VPN

Module 1　Text Study 课文学习

Basic Training 基本训练

Text 1　根据语音和视频完成以下任务。

Task 1-0 听录音，记录关键词，理解课文大意。

_____
_____
_____
_____
_____

L11-1-1.MP3

L11-1-2.MP3

L11-1.MP4

Task 1-1 翻译画线的部分。

① 用物理层设备连接的网络一般称为一个网段_____，也可以泛指子网_____，例如，用子网掩码_____划分的子网。

② 虚拟的（virtual）和实际的（_____）对应，逻辑的（_____）和物理的（_____）对应。

Task 1-2 英汉互译。

带宽分配_____　　　资源优化_____　　　基于端口的_____
终端站点_____　　　成员身份_____　　　负载均衡_____
picture_____　　　map_____

Task 1-3 填空。

keep track of 和_____基本是一个意思，表示跟踪、记录痕迹。

Task 1-4 翻译。

① Virtual LAN (VLAN) refers to a group of logically networked devices on one or more LANs that are configured so that they can communicate as if they were attached to the same wire, when in fact they are located on a number of different LAN segments.

② The virtual LAN controller can change or add workstations and manage load-balancing and bandwidth allocation more easily than with a physical picture of the LAN.

③ Each physical switch port is configured with an access list specifying membership in a set of VLANs.

④ Network management software keeps track of relating the virtual picture of the Local Area Network with the actual physical picture.

课文

VLAN

Virtual LAN (VLAN, Figure 2-11) refers to a group of logically networked devices on one or more LANs that are configured so that they can communicate as if they were attached to the same wire, when in fact they are located on a number of different LAN segments. Because VLANs are based on logical connections instead of physical connections，it is very flexible for user/host management, bandwidth allocation and resource optimization.

A virtual (or logical) LAN is a Local Area Network with a definition that maps workstations on some other basis than geographic location (for example, by department, type of user, or primary application). The virtual LAN controller can change or add workstations and manage load-balancing and bandwidth allocation more easily than with a physical picture of the LAN. Network management software keeps track of relating the virtual picture of the Local Area Network with the actual physical picture.

There are some types of Virtual LANs as the following:

● Port-Based VLAN: Each physical switch port is configured with an access list specifying membership in a set of VLANs.

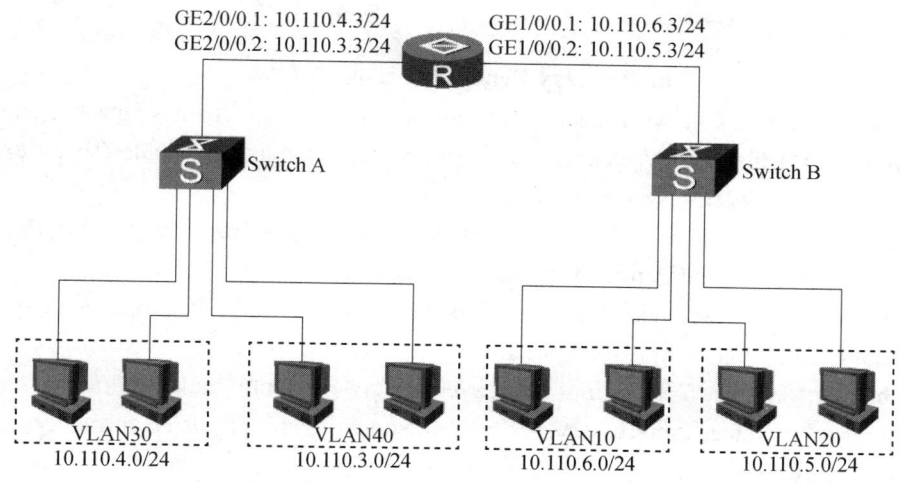

GE2/0/0.1: 10.110.4.3/24   GE1/0/0.1: 10.110.6.3/24
GE2/0/0.2: 10.110.3.3/24   GE1/0/0.2: 10.110.5.3/24

Switch A   Switch B

VLAN30   VLAN40   VLAN10   VLAN20
10.110.4.0/24   10.110.3.0/24   10.110.6.0/24   10.110.5.0/24

Figure 2-11   VLAN

● MAC-Based VLAN: A switch is configured with an access list mapping individual MAC addresses to VLAN membership.

● Protocol-Based VLAN: A switch is configured with a list of mapping layer 3 protocol types to VLAN membership—thereby filtering IP traffic from nearby end-stations using a particular protocol such as IPX.

VLANs are considered likely to be used with campus environment networks. Among companies likely to provide products with VLAN support are Cisco, Bay Networks, H3C，Ruijie and TP-LINK.

There is a proposed VLAN standard, IEEE (Institute of Electrical and Electronics Engineers) 802.10.

Text 2　根据语音和视频完成以下任务。

Task 2-0 听录音，记录关键词，理解课文大意。

_____

_____

L11-2.MP3

_____

Take 2-1 写出主要的中文含义，并指出在本课文或其他课文中增加或减少的对象是什么？

L11-2.MP4

① decrease_____

② reduce/reduction_____

③ mitigate/mitigation_____

④ boost_____

⑤ improve/improvement_____

⑥ increase_____

Task 2-2 词汇辨析。

allocate 和 assign_____

课文

The Primary Benefits of Using VLANs

User productivity and network adaptability are key drivers for business growth and success. Implementing VLAN technology enables a network to support business goals more flexibly. The primary benefits of using VLANs are as follows:

- Security—Groups that have sensitive data are separated from the rest of the network, decreasing the chances of confidential information breaches.
- Cost reduction—Cost savings result from less need for expensive network upgrades and more efficient use of existing bandwidth and uplinks.
- Higher performance—Dividing flat Layer 2 networks into multiple logical workgroups (broadcast domains) reduces unnecessary traffic on the network and boosts performance.
- Broadcast storm mitigation—Dividing a network into VLANs reduces the number of devices that may participate in a broadcast storm.
- Improved IT staff efficiency—VLANs make it easier to manage the network because users with similar network requirements share the same VLAN. When you provision a new switch, all the policies and procedures already configured for the particular VLAN are implemented when the ports are assigned. It is also easy for the IT staff to identify the function of a VLAN by giving it an appropriate name.
- Simpler project or application management—VLANs aggregate users and network devices to support business or geographic requirements, for having separate functions makes managing a project or working with a specialized application easier. It is also easier to determine the scope of the effects of upgrading network services.

Advanced Training 进阶训练

T11.PDF

根据下文完成以下任务。

Task 1 翻译。

① VPNs are often used to extend Intranets worldwide to disseminate information and news to a wide user base.

② IPSec is often referred to as a "security overlay" because of its use as a security layer for other protocols.

③ PPTP is one of the most widely used VPN protocols because of its straightforward configuration and maintenance and also because it is included in the Windows operating system.

④ VPN technology employs sophisticated encryption to ensure security and prevent any unintentional interception of data between private sites.

⑤ The use of the Internet as the main communications channel between sites is a cost-effective alternative to expensive leased private lines.

⑥ The relative ease, speed, and flexibility of VPN provisioning in comparison to leased lines makes VPNs an ideal choice for corporations who require flexibility.

Task 2 词义辨析。
private（私有的）和 confidential/secret（机密的）有什么区别？

Task 3 PIN 是什么的缩写？ _____

VPN

A Virtual Private Network (VPN, Figure 2-12) is a network technology that creates a secure network connection over a public network such as the Internet or a private network owned by service providers. Large corporations, educational institutions, and government agencies use VPN technology to enable remote users to securely connect to a private network.

A VPN can connect multiple sites over a large distance just like a Wide Area Network (WAN). VPNs are often used to extend Intranets worldwide to disseminate information and news to a wide user base. Educational institutions use VPNs to connect campuses that can be distributed across the country or around the world.

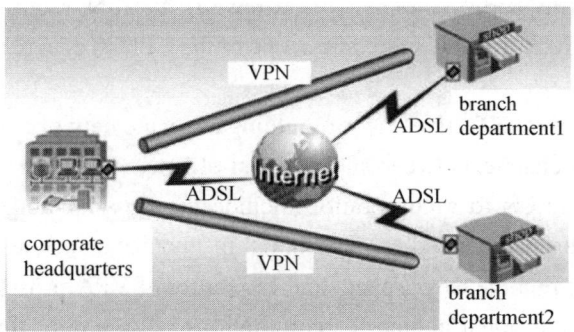

Figure 2-12　VPN

In order to gain access to the private network, a user must be authenticated using a unique identification and a password. An authentication token is often used to gain access to a private network through a Personal Identification Number (PIN) that a user must enter. The PIN is a unique authentication code that changes according to a specific frequency, usually every 30 seconds or so.

There are a number of VPN protocols in use that secure the transport of data traffic over a public network infrastructure. Each protocol varies slightly in the way that data is kept secure.

IP Security (IPSec) is used to secure communications over the Internet. IPSec traffic can use either transport mode or tunneling to encrypt data traffic in a VPN. The difference between the two modes is that transport mode encrypts only the message within the data packet (also known as the payload) while tunneling encrypts the entire data packet. IPSec is often referred to as a "security overlay" because of its use as a security layer for other protocols.

Secure Sockets Layer (SSL) and Transport Layer Security (TLS) use cryptography to secure communications over the Internet. Both protocols use a "handshake" method of authentication that involves a negotiation of network parameters between the client and server machines. To successfully initiate a connection, an authentication process involving certificates is used. Certificates are cryptographic keys that are stored on both the server and client. Point-to-Point Tunneling Protocol (PPTP) is another Tunneling Protocol used to connect a remote client to a private server over the Internet. PPTP is one of the most widely used VPN protocols because of its straightforward configuration and maintenance and also because it is included in the Windows operating system.

Layer 2 Tunneling Protocol (L2TP) is a protocol used to tunnel data communications traffic between two sites over the Internet. L2TP is often used in tandem with IPSec (which acts as a security layer) to secure the transfer of L2TP data packets over the Internet. Unlike PPTP, a VPN implementation using L2TP/IPSec requires a shared key or the use of certificates.

VPN technology employs sophisticated encryption to ensure security and prevent any unintentional interception of data between private sites. All traffic over a VPN is encrypted using algorithms to secure data integrity and privacy. VPN architecture is governed by a strict set of rules and standards to ensure a private communication channel between sites. Corporate network administrators are responsible for deciding the scope of a VPN, implementing and deploying a VPN, and constantly monitoring network traffic across the network firewall. A VPN requires administrators to be continually aware of the overall architecture and scope of the VPN to ensure communications are kept private.

A VPN is an inexpensive effective way of building a private network. The use of the Internet as the main communications channel between sites is a cost-effective（划算的） alternative to expensive leased private lines. The costs to a corporation include the network authentication hardware and software used to authenticate users and any additional mechanisms such as authentication tokens or other secure devices. The relative ease, speed, and flexibility of VPN provisioning in comparison to leased lines makes VPNs an ideal choice for corporations who require flexibility. For example, a company can adjust the number of sites in the VPN according to changing requirements.

There are several potential disadvantages with VPN use. The lack of Quality of Service (QoS) management over the Internet can cause packet loss and other performance issues. Adverse network conditions that occur outside of the private network are beyond the control of the VPN administrator. For this reason, many large corporations pay for the use of trusted VPNs that use a private network to guarantee QoS. Vendor interoperability is another potential disadvantage as VPN technologies from one vendor may not be compatible with VPN technologies from another vendor. Neither of these disadvantages has prevented the widespread acceptance and deployment of VPN technology.

Vocabulary Practice 词汇练习

词汇及短语听写，并纠正发音。

C11.MP3

Words & Expressions

optimization [ˌɒptɪmaɪˈzeɪʃən]	*n.* 最佳化，最优化；优化组合
confidential [ˌkɒnfɪˈdenʃl]	*adj.* 秘密的；机密的
breach [briːtʃ]	*n.* 破坏；破裂；违背
boost [buːst]	*vt.* 促进，提高；使增长；使兴旺
mitigation [ˌmɪtɪˈgeɪʃn]	*n.* 缓解，减轻，平静
provision [prəˈvɪʒn]	*v.* 为……提供所需物品
certificate [səˈtɪfɪkət]	*n.* 证明；结业证书
in tandem [ɪnˈtændəm]	*adv.* 一前一后地；协力地；并驾齐驱；同时实行
sophisticated [səˈfɪstɪkeɪtɪd]	*adj.* 复杂的；精致的；富有经验的；深奥微妙的
unintentional [ˌʌnɪnˈtenʃənl]	*adj.* 不是故意的；无意的，无心的
interception [ˌɪntəˈsepʃn]	*n.* 拦截；拦阻；截断
adverse [ˈædvɜːs]	*adj.* 不利的；有害的
interoperability [ˈɪntərˌɒpərəˈbɪlɪtɪ]	*n.* 互用性，协同工作的能力

Module 2   Study Aid 辅助帮学

Terms & Abbreviations 术语和缩写

VLAN   *abbr.* 虚拟局域网（Virtual LAN），是一组逻辑上的设备和用户，这些设备和用户并不受物理位置的限制，可以根据功能、部门及应用等因素分类、组织，相互之间的通信就好像它们在同一个局域网中一样，而 VLAN 之间的通信可以通过第三层的路由来完成。

LAN segment   局域网段，指通过物理层设备（传输介质、中继器、集线器等）互联，能够直接通信的那一部分网络是以太网上的一个广播域。而网段（segment）一般是指网络层的子网，例如，从 192.168.0.1 到 192.168.0.255，当子网掩码为 255.255.255.0 时是一个网段。

IPX   *abbr.* 互联网分组交换（Internetwork Packet eXchange），是一个专用的协议簇，主要由 Novell NetWare 操作系统使用。IPX 是 IPX 协议簇中的第三层协议，用于将数据从服务器发送到工作站。

broadcast storm   广播风暴，指在有环路的网络中，如果数据帧始终有效，同样的数据帧就会被 hub/switch 持续、反复复制和传播，网络带宽被迅速占用，设备满负荷工作，直至网络完全瘫痪。

VPN   *abbr.* 虚拟专用网（Virtual Private Network），指在公用网络上通过加密通信建立的专用网络，在企业网络中有广泛应用。VPN 网关通过对数据包的加密和数据包目标地址的转换实现远程访问。

PIN   *abbr.* 个人标识号（Personal Identification Number），用于确保智能卡免受误用的秘密标识代码。PIN 与密码类似，只有卡的所有者才知道该 PIN 的内容。

IPSec   *abbr.* 互联网协议安全（IP Security），一种开放标准的框架结构，通过使用加密的安全服务以确保在 Internet 协议（IP）网络上进行保密而安全的通信。

payload　　　有效负载，是用户需要传输的那部分原始数据。在传输数据时，为了使数据传输可行和可靠，需要把原始数据分批传输，并且在每一批数据的头和尾都加上一定的辅助信息，比如，这一批数据的大小、校验位等。

SSL　　*abbr.* 安全套接层（Secure Sockets Layer），是为网络通信提供安全及数据完整性的一种安全协议，在传输层对网络连接进行加密。

TLS　　*abbr.* 传输层安全（Transport Layer Security），是为网络通信提供安全及数据完整性的一种安全协议。TLS 与 SSL 在传输层对网络连接进行加密。

PPTP　　*abbr.* 点到点隧道协议（Point-to-Point Tunneling Protocol）。该协议是在 PPP 的基础上开发的一种新的增强型安全协议，支持多协议虚拟专用网（VPN），可以通过密码验证协议（PAP）、可扩展认证协议（EAP）等增强安全性。

L2TP　　*abbr.* 第二层隧道协议（Layer 2 Tunneling Protocol），是一种工业标准的 Internet 隧道协议，功能大致和 PPTP 类似，比如，同样可以对网络数据流进行加密。不过也有不同之处，比如，PPTP 要求网络为 IP 网络，L2TP 要求面向数据包的点对点连接；PPTP 使用单一隧道，L2TP 使用多隧道；L2TP 提供包头压缩、隧道验证，而 PPTP 不支持。

QoS　　*abbr.* 服务质量（Quality of Service），指一个网络能够利用各种技术，为指定的网络通信提供指定服务（如带宽、延时等）的能力。

Difficult Sentences Analysis and Translation 难句分析与翻译

1. Virtual LAN (VLAN) refers to a group of logically networked devices on one or more LANs that are configured so that they can communicate as if they were attached to the same wire, when in fact they are located on a number of different LAN segments.

译文：虚拟局域网（VLAN）是指在一个或多个局域网上的一组逻辑上的网络设备。这些设备的配置是为了能够通信，就好像它们是连接在同一根线缆上的，然而，事实上它们位于多个不同的局域网段上。

分析：句中"that"引导的定语从句修饰、限定"a group of logically networked devices"，其中包含"so that"引导的目的状语从句和"as if"引导的虚拟语气条件句；"when in fact they are located on a number of different LAN segments"是状语从句，其中"when"表示对比，相当于"whereas"，译为"然而"。

2. VLANs aggregate users and network devices to support business or geographic requirements, for having separate functions makes managing a project or working with a specialized application easier.

译文：虚拟局域网聚合用户和网络设备以支持业务或地理的要求，因为虚拟局域网具有独立的功能，从而使管理项目或使用专门的应用程序更为容易。

分析：句中连词"for"引导原因状语从句，提供辅助性的补充说明；动名词短语"having separate functions"在从句中充当主语；动名词短语"managing a project or working with a specialized application"在从句中充当动词"makes"的宾语，easier 为其宾语补足语。

3. A Virtual Private Network (VPN) is a network technology that creates a secure network connection over a public network such as the Internet or a private network owned by service providers.

译文：虚拟专用网（VPN）是一种通过诸如互联网或服务提供商拥有的私有网络这样的公共网络来创建安全网络连接的网络技术。

分析：句中连词"that"引导定语从句，修饰、限定前面的"a network technology"。

4. The difference between the two modes is that transport mode encrypts only the message within the data packet (also known as the payload) while tunneling encrypts the entire data packet.

译文：这两种模式之间的区别在于传输模式只对数据包中的消息（也称为有效负载）进行加密，而隧道加密对整个数据包进行加密。

分析：句中连词"that"引导表语从句；在表语从句中"while"用作连词，连接两个并列句子"transport mode encrypts only the message within the data packet (also known as the payload)"和"tunneling encrypts the entire data packet"，表示对比关系。

5. PPTP is one of the most widely used VPN protocols because of its straightforward configuration and maintenance and also because it is included in the Windows operating system.

译文：点到点隧道协议是使用最广泛的虚拟专用网协议之一，不仅因为它配置和维护简单，而且因为它就包含在 Windows 操作系统中。

分析：句中连词"because of"后接名词短语，表示原因；"because it is included with the Windows operating system"是原因状语从句。两个原因是并列关系，由并列连词"and"连接。

6. Corporate network administrators are responsible for deciding the scope of a VPN, implementing and deploying a VPN, and constantly monitoring network traffic across the network firewall.

译文：公司网络管理员负责决定虚拟专用网的范围，实施和部署虚拟专用网，并不断监控网络防火墙的网络流量。

分析：句中动名词短语"deciding the scope of a VPN""implementing and deploying a VPN"和"monitoring network traffic across the network firewall"都是"are responsible for"的介词宾语。

7. Vendor interoperability is another potential disadvantage as VPN technologies from one vendor may not be compatible with VPN technologies from another vendor.

译文：由于来自一个供应商的虚拟专用网技术可能与另一个供应商的虚拟专用网技术不兼容，供应商的互操作性则是另一个潜在的缺点。

分析：句中"as"引导原因状语从句，表示原因比较明显；"be compatible with"意思是"与……兼容"。

Module 3   Consolidation Exercise 巩固练习

K11.PDF

I. Give the meaning of the following abbreviated terms both in English and in Chinese

VLAN	VPN	PIN	IPSec
SSL	TLS	PPTP	L2TP
QoS			

II. Answer the following questions according to the passage

1. Why are VLANs very flexible for user/host management, bandwidth allocation and resource

optimization?

2. How many types of Virtual LANs are mentioned in Text 1? What are they?

3. What is proposed as VLAN standard?

4. What are the primary benefits of using VLANs?

5. Why do large corporations, educational institutions, and government agencies use VPN?

6. Does the PIN change very frequently during the communication?

7. What VPN protocols are in use?

8. What is the difference between transport mode and tunneling?

9. How does "handshake" method work?

10. What are the potential disadvantages of VPN?

III. Fill in each blank with appropriate words or expressions according to the passage

1. Virtual LAN (VLAN) refers to a group of _____ networked devices on one or more LANs that are _____ so that they can communicate as if they were attached to the same _____, when in fact they are located on a number of different LAN _____.

2. A virtual LAN is _____ with a definition that maps workstations on some other basis than _____ location. The virtual LAN controller can change or add _____ and manage _____ and bandwidth allocation more easily than with a physical picture of the LAN.

3. A Virtual Private Network (VPN) is a network technology that creates a _____ network connection over a _____ such as the Internet or a private network owned by service _____.

4. In order to gain access to the private network, a user must be _____ using a unique identification and a password. An authentication _____ is often used to gain access to a private network through _____ that a user must enter.

5. _____ and _____ use cryptography to secure communications over the Internet. Both protocols use a "handshake" method of authentication that involves a negotiation of network parameters between the _____ machines.

6. Layer 2 Tunneling Protocol (L2TP) is a protocol used to tunnel _____ between two sites over the Internet. L2TP is often used in tandem with IPSec (which acts as _____) to secure the transfer of L2TP _____ over the Internet.

7. The relative ease, speed, and _____ of VPN provisioning in comparison to _____ lines makes VPNs an ideal choice for _____ who require flexibility. For example, a company can _____ the number of sites in the VPN according to changing _____.

8. Vendor _____ is another potential disadvantage as VPN technologies from one vendor may not _____ VPN technologies from another vendor.

Lesson 12   NOS and Virtual Machine

Module 1   Text Study 课文学习

Basic Training 基本训练

Text 1   根据语音和视频完成以下任务。

Task 1-0 听录音，记录关键词，理解课文大意。

_____

_____

L12-1-1.MP3

_____

_____

L12-1-2.MP3

_____

_____

Task 1-1 用英文填空。

① operate the functions 可以直接用_____的动词形式来表达，但"有多项功能"的意思就无法表达出来。

L12-1.MP4

② 文中"A specialized operating system…Examples: JUNOS, used in routers and switches from Juniper Networks, Cisco IOS (formerly "Cisco Internetwork Operating System")."

JUNOS 和 Cisco IOS 都是_____的例子。Juniper Networks（瞻博网络）和

Cisco（思科）是两家著名的网络公司。

③ 和 virtual（虚拟的）对应的是_____、_____和_____，例如，vlan 和_____ network，VM 和_____ machine 或_____ machine。_____（真实的，实际的；现行的，目前的）指已经发生或存在的事物，来源于"行动的结果"；_____（实际的；真实的；实在的）强调不是理想或虚幻，表示"本来就是"，来源于"真、实"。

Task 1-2 写出对应的英文表达。

虚拟现实_____　　虚拟机实例_____

模拟一个架构_____　　废弃的平台_____

可移植性_____　　灵活性_____

硬件虚拟化_____　　成本效益_____

Task 1-3 按课文中的观点，NOS 和 common operating systems 目前有区别吗？

_____

Task 1-4 分别用 divide 和 separate 翻译下面两句话。

① 网络操作系统分为网络设备专用操作系统和面向计算机网络的操作系统。

_____

② 根据与真实机器相似的程度，虚拟机主要分为系统虚拟机和进程虚拟机，即语言虚拟机。

_____

Task 1-5 翻译。

① A Virtual Machine (VM) is a software implementation of a machine (for example, a computer) that executes programs like a physical machine.

_____

② This sense is now largely historical, as common operating systems generally now have such features included.

_____

课文

NOS

The term Network Operating System is used to refer to two rather different concepts:

A specialized operating system for a network device such as a router, switch or firewall that operates the functions in the network layer (layer 3). Examples: JUNOS, used in routers and switches from Juniper Networks, Cisco IOS (formerly "Cisco Internetwork Operating System").

An operating system oriented to computer networking, which allows shared file and printer access among multiple computers in a network, and which enables the sharing of data, users, groups, security, applications, and other networking functions, typically over a Local Area Network (LAN) or private network. This sense is now largely historical, as common operating systems generally now have such features included.

VM

A Virtual Machine (VM, Figure 2-13) is a software implementation of a machine (for example, a computer) that executes programs like a physical machine. Virtual Machines are separated into two major classes, based on their use and degree of correspondence to any real machine.

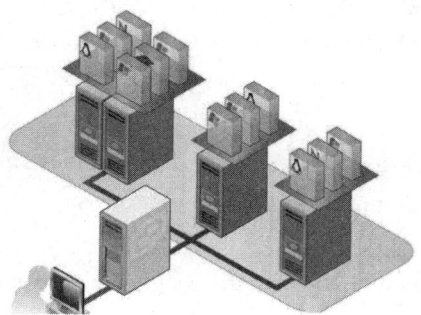

Figure 2-13   Virtual Machine

A system Virtual Machine provides a complete system platform which supports the execution of a complete Operating System (OS). These usually emulate an existing architecture, and are built with the purpose of either providing a platform to run programs where the real hardware is not available for use (for example, executing on otherwise obsolete platforms), or having multiple instances of Virtual Machines leading to more efficient use of computing resources, both in terms of energy consumption and cost-effectiveness (known as hardware virtualization, the key to a cloud computing environment), or both.

A Process Virtual Machine (also, Language Virtual Machine) is designed to run a single program, which means that it supports a single process. Such Virtual Machines are usually closely suited to one or more programming languages and built with the purpose of providing program portability and flexibility amongst other things. An essential characteristic of a Virtual Machine is that the software running inside is limited to the resources and other abstractions provided by the Virtual Machine—it cannot break out of its virtual environment.

Text 2   **根据语音和视频完成以下任务。**

Task 2-0 听录音，记录关键词，理解课文大意。

_____

_____   L12-2-1.MP3

_____

_____   L12-2-2.MP3

_____

_____

Task 2-1 英汉互译。   L12-2.MP4

类 Unix 家族_____   桌面操作系统_____   mainframe_____

稳定的性质_____   开源软件_____   底层源代码_____

专用的软件_____   内核_____   重新发布_____

许可证_____   库_____   工具_____或_____

基金会_____   派生物_____   游戏主机_____

Task 2-2 填空。

① GNU 是＿＿＿＿＿＿＿＿＿＿＿的递归首字母缩写词，发音为 g'noo。

② ＿＿＿＿＿＿＿＿＿＿的中文名是"乌班图"，但并不是所有的英文命名都有对应的中文名，例如，Debian 没有中文名。根据 Debian 的官方网站建议，读作：Deb'-ee-en，谐音为"得比恩"，在译文中可以使用原英文名。

Task 2-3 翻译。

① Solaris supports SPARC-based and x86-based workstations and servers from Oracle and other vendors, with efforts underway to port to additional platforms.

＿＿

② Solaris is registered as compliant with the Single UNIX Specification.

＿＿

③ Other distributions may include a less resource intensive desktop such as LXDE or Xfce for use on older or less-powerful computers.

＿＿

Task 2-4

Fedora、SUSE、KDE、Plasma、Sun Microsystems 等词如何发音？如何翻译？

＿＿

Task 2-5 分别用下列词汇造句。

① extremely＿＿＿＿＿＿＿＿＿＿＿＿＿＿＿＿＿＿＿＿＿＿＿＿＿＿＿＿＿＿＿＿＿＿

② especially＿＿＿＿＿＿＿＿＿＿＿＿＿＿＿＿＿＿＿＿＿＿＿＿＿＿＿＿＿＿＿＿＿＿

③ innovative＿＿＿＿＿＿＿＿＿＿＿＿＿＿＿＿＿＿＿＿＿＿＿＿＿＿＿＿＿＿＿＿＿＿

④ prominent＿＿＿＿＿＿＿＿＿＿＿＿＿＿＿＿＿＿＿＿＿＿＿＿＿＿＿＿＿＿＿＿＿＿

⑤ typically＿＿＿＿＿＿＿＿＿＿＿＿＿＿＿＿＿＿＿＿＿＿＿＿＿＿＿＿＿＿＿＿＿＿＿

⑥ commonly＿＿＿＿＿＿＿＿＿＿＿＿＿＿＿＿＿＿＿＿＿＿＿＿＿＿＿＿＿＿＿＿＿＿

课文

UNIX

As one of the common operating systems, Linux refers to the family of Unix-like computer operating systems using the Linux kernel. Linux can be installed on a wide variety of computer hardware, ranging from mobile phones, tablet computers and video game consoles, to mainframes and supercomputers. Linux is a leading server operating system, and runs the 10 fastest supercomputers in the world. It is also a top contender in the desktop OS market, due to its extremely secure and stable nature, its speed, and its lack of fragmentation issues. The development of Linux is one of the most prominent examples of free and open-source software collaboration. Typically, all the underlying source code can be used, freely modified, and redistributed, both commercially and non-commercially, by anyone under licenses such as the GNU GPL (General Public License). Typically, Linux is packaged in a format known as a Linux distribution for desktop and server use. Some popular mainstream Linux distributions include Debian (and its derivatives such as Ubuntu), Fedora and open SUSE. Linux distributions include the Linux kernel and supporting utilities and libraries to fulfill the distribution's intended use.

A distribution oriented toward desktop use may include the X Window System, the GNOME and KDE Plasma desktop environments. Other distributions may include a less resource intensive desktop such as LXDE or Xfce for use on older or less-powerful computers. Because Linux is freely redistributable, it is possible for anyone to create a distribution for any intended use. Commonly used applications with desktop Linux systems include the Mozilla Firefox Web browser, the OpenOffice.org office application suite and the GIMP image editor.

The name "Linux" comes from the Linux kernel, originally written in 1991 by Linus Torvalds. The main supporting user space system tools and libraries from the GNU Project (announced in 1983 by Richard Stallman) are the basis for the Free Software Foundation's preferred name GNU/Linux.

Solaris is a Unix operating system originally developed by Sun Microsystems. Solaris is known for its scalability, especially on SPARC systems, and for originating many innovative features such as DTrace, ZFS and Time Slider. Solaris supports SPARC-based and x86-based workstations and servers from Oracle and other vendors, with efforts underway to port to additional platforms. Solaris is registered as compliant with the Single UNIX Specification. Historically, Solaris was developed as proprietary software.

Advanced Training 进阶训练

T12.PDF

根据课文完成以下任务。

Task 1 翻译。

① As such it is not optimized for use as a file server or media server.

_____

② NetWare 6.5SP8 General Support ended in 2010, with Extended Support until the end of 2015, and Self Support until the end of 2017.

_____

Task 2 回答问题。

What do Win2K8, Inc stand for respectively? How to pronounce them?

_____

Task 3 翻译。

即插即用_____   热插拔的_____

BitLocker_____   Itanium-based_____

IA-64_____

Task 4 回答问题。

managerial 和 administrative 在词义上有什么区别？文中如何翻译？

_____

Examples of Other Common Operating Systems

Windows Server 2008 (sometimes abbreviated as "Win2K8") is one of Microsoft Windows' server operating systems. Like Windows Vista and Windows 7, Windows Server 2008 is built on Windows NT 6.x. Therefore, it shares much of the same architecture and functionality. Since the code base is common, it automatically comes with most of the technical, security, managerial and administrative features new to Windows Vista such as the rewritten networking stack (native IPv6, native wireless, speed and security improvements); improved image-based installation, deployment and recovery; improved diagnostics, monitoring, event logging and reporting tools; new security features such as BitLocker and ASLR; improved Windows Firewall with secure default configuration; NET Framework 3.0 technologies, specifically WCF (Windows Communication Foundation), MSMQ (Microsoft Message Queuing) and WWF (Windows Workflow Foundation); and the core kernel, memory and file system improvements. Processors and memory devices are modeled as PnP (Plug and Play) devices, to allow hot-plugging of these devices. This allows the system resources to be partitioned dynamically using Dynamic Hardware Partitioning, each partition having its own memory, processor and I/O host bridge devices independent of other partitions. Most editions of Windows Server 2008 are available in x86-64 (64-bit) and x86 (32-bit) versions. Windows Server 2008 for Itanium-based Systems supports IA-64 processors. Microsoft has optimized the IA-64 version for high-workload scenarios like database servers and Line of Business (LOB) applications. As such it is not optimized for use as a file server or media server. Microsoft has announced that Windows Server 2008 is the last 32-bit Windows server operating system. Windows Server 2008 is available in some editions, similar to Windows Server 2003.

NetWare is a computer Network Operating System developed by Novell, Inc. It initially used cooperative multitasking to run various services on a personal computer, using the IPX network protocol. The final update release was version 6.5SP8 of May 2009. Netware is no longer on Novell's product list. NetWare 6.5SP8 General Support ended in 2010, with Extended Support until the end of 2015, and Self Support until the end of 2017. The replacement is Open Enterprise Server.

Vocabulary Practice 词汇练习

词汇及短语听写，并纠正发音。

C12.MP3

_____

_____

_____

_____

Words & Expressions

queue [kju:]	vi. 排队等候　vt.（使）排队，列队等待
partition [pɑːˈtɪʃn]	vt. 分隔开；区分；分割　n. 分区；划分，分开
console [kənˈsəʊl]	n. 控制台，操纵台
mainframe [ˈmeɪnfreɪm]	n.（大型电脑的）主机；总配线架
contender [kənˈtendə(r)]	n.（冠军）争夺者，竞争者
fragmentation [ˌfrægmenˈteɪʃn]	n. 破碎，磁盘碎片
prominent [ˈprɒmɪnənt]	adj. 突出的，杰出的；突起的；著名的
underlying [ˌʌndəˈlaɪɪŋ]	adj. 潜在的，基础的；根本的；表面下的
derivative [dɪˈrɪvətɪv]	n. 派生物；衍生物；派生词；衍生字
redistributable [ˌriːdɪstrɪˈbjʊtəbl]	adj. 可再分配的，可再发行的
innovative [ˈɪnəveɪtɪv]	adj. 革新的；创新的

compliant [kəmˈplaɪənt]　　　　　*adj.* 遵从的；依从的；符合的；一致的
emulate [ˈemjuleɪt]　　　　　　　*vt.* 仿真，模仿；同……竞争；努力赶上
obsolete [ˈɒbsəliːt]　　　　　　　*adj.* 废弃的；老式的，已过时的
portability [ˌpɔːtəˈbɪlətɪ]　　　　*n.* 便携性，轻便

Module 2　Study Aid 辅助帮学

Terms & Abbreviations 术语和缩写

BitLocker　　　一种驱动器加密技术，加密 Windows 操作系统，可以很好地保护所有数据，确保计算机在无人参与、丢失或被盗的情况下也不会被访问。

ASLR　　　*abbr.* 地址空间配置随机加载（Address Space Layout Randomization），是一种针对缓冲区溢出的安全保护技术，如今 Linux、FreeBSD、Windows 等主流操作系统都已采用了该技术。

NET Framework　　　NET 框架，是 Windows 的新托管代码编程模型。它将强大的功能与新技术结合起来，用于构建具有视觉上引人注目的用户体验的应用程序，实现跨技术边界的无缝通信，并且能支持各种业务流程。

WCF　　　*abbr.* Windows 通信基础（Windows Communication Foundation），是由 Microsoft 开发的一系列支持数据通信的应用程序框架。它是 Microsoft 为构建面向服务的应用程序而提供的统一编程模型，它以一种新型的以服务为向导的编程模型来简化应用程序开发，对安全、可靠的数据交换，以及多种方式的传输选择和各种编码设置都提供了便利的支持。

MSMQ　　　*abbr.* 微软消息队列（Microsoft Message Queuing），应用程序开发人员可以通过发送和接收消息，方便地与应用程序进行快速、可靠的通信，是在多个不同的应用之间实现相互通信的一种异步传输模式，相互通信的应用可以分布于同一台机器上，也可以分布于相连的网络空间中的任意位置。

WWF　　　*abbr.* Windows 工作流基础（Windows Workflow Foundation），是一个可扩展框架，用于在 Windows 平台上开发工作流解决方案。

kernel　　　内核，操作系统大致可以分为内核（kernel）和外壳（shell），内核负责硬、软件资源管理相关事务，而外壳处理与用户使用相关的事务。

PnP　　　*abbr.* 即插即用（Plug and Play），其作用是自动配置计算机中的板卡和其他设备，告诉对应的设备做了什么。

hot-plug　　　热插拔，指带电插拔。

Itanium　　　Intel 安腾处理器，构建在 IA-64（Intel Architecture 64）上。与 x86 不同，它专门用在高端企业级 64-bit 计算环境中，但所有基于安腾处理器的系统都支持 32Intel 架构（IA-32）的软件应用，从而为用户移植到 Intel 安腾架构提供了更高的灵活性。

GNU　　　革奴计划，是 "GNU is Not Unix" 的首字母缩写，它的目标是创建一套完全自由的操作系统。为保证 GNU 软件可以自由地使用、复制、修改和发布，所有 GNU 软件都有一份在禁止其他人添加任何限制的情况下授权所有权利给任何人的协议条款。

GPL　　　*abbr.* 通用公共许可证（General Public License），是一个广泛被使用的自由软件许可证。

GNOME　　*abbr.* GNU 网络对象模型环境（GNU Network Object Model Environment），是一套纯粹的、自由的计算机软件，运行在操作系统上，提供图形桌面环境。

KDE　　*abbr.* K 桌面环境（K Desktop Environment），是一种著名的运行于 Linux、Unix 及 FreeBSD 等操作系统上的自由图形桌面环境。

LXDE　　LXDE 桌面环境，是一个基于 GTK2 的美观和国际化的桌面环境。跟 KDE 和 GNOME 比起来，LXDE 占用的资源更少，适合在配置比较差的计算机上工作。

Xfce　　Xfce 桌面环境，是一个轻量级的类 Unix 的桌面系统，这个词的发音为 X-f-c-e（即四个字母一个一个地读）。

SPARC　　*abbr.* 可扩充处理器架构（Scalable Processor Architecture），是 RISC 微处理器架构之一。它由 Sun 设计，也是公司注册商标之一。

Oracle　　甲骨文股份有限公司，是全球最大的数据库软件公司。

VM　　*abbr.* 虚拟机（Virtual Machine），是指通过软件模拟的具有完整硬件系统功能、运行在一个完全隔离环境中的完整计算机系统。

IA-64　　英特尔安腾架构（Intel Itanium architecture）64 位元指令集，此架构与 x86 及 x86-64 并不相容，操作系统与软件需使用 IA-64 专用版本。

Difficult Sentences Analysis and Translation 难句分析与翻译

1. An operating system oriented to computer networking, which allows shared file and printer access among multiple computers in a network, and which enables the sharing of data, users, groups, security, applications, and other networking functions, typically over a Local Area Network (LAN) or private network.

译文：一种面向计算机网络的操作系统，允许网络中的多台计算机之间共享文件和打印机，以便共享数据、用户、组、安全、应用程序和其他网络功能，通常是在局域网（LAN）或专用网络上进行的。

分析：名词性短语"An operating system oriented to computer networking"是计算机网络操作系统的概念之一；"which allows shared file…in a network"和"which enables the sharing of data,… or private network"是非限制性定语从句，其中的关系代词"which"指代前面的名词性短语"An operating system oriented to computer networking"，在从句中作主语。

2. Since the code base is common, it automatically comes with most of the technical, security, management and administrative features new to Windows Vista such as the rewritten networking stack (native IPv6, native wireless, speed and security improvements); improved image-based installation, deployment and recovery; improved diagnostics, monitoring, event logging and reporting tools; new security features such as BitLocker and ASLR; improved Windows Firewall with secure default configuration; NET Framework 3.0 technologies, specifically WCF (Windows Communication Foundation), MSMQ (Microsoft Message Queuing) and WWF (Windows Workflow Foundation); and the core kernel, memory and file system improvements.

译文：由于基础代码是一样的，它自动附带了 Windows Vista 新功能的大部分技术、安全、管理和管理特征，例如：重写的网络堆栈（自带的 IPv6、自带的无线、速度和安全性的改进）；改进的基于图像的安装、部署和恢复；改进的诊断、监测、事件记录和报告工具；新的安全功能，如 BitLocker 驱动器加密和地址空间配置随机加载 ASLR；改进的具有安全默认配置的

Windows 防火墙；NET 框架 3.0 技术，特别是 Windows 通信基础、微软消息队列和 Windows 工作流基础；还有核心内核、内存和文件系统的改进。

分析：本句句子长、术语较多，但句子结构并不复杂。句中"Since"引导原因状语从句，意思是"既然、由于"，表示原因已知；主句中"such as"后的名词性短语用分号隔开，均为并列关系；名词性短语"new security features"后的"such as"引出实例。

3. These usually emulate an existing architecture, and are built with the purpose of either providing a platform to run programs where the real hardware is not available for use (for example, executing on otherwise obsolete platforms), or having multiple instances of Virtual Machines leading to more efficient use of computing resources, both in terms of energy consumption and cost-effectiveness (known as hardware virtualization, the key to a cloud computing environment), or both.

译文：这些虚拟机通常模拟现有的体系结构，其目的要么是提供一个平台在实际硬件不可用时来运行程序（例如，在原本过时的平台上执行程序），要么在能源消耗和成本效益（称为硬件虚拟化，这是云计算环境的关键）方面，有多个虚拟机实例来更有效地利用计算资源，或者两者兼而有之。

分析：句中"These"指上句提到的虚拟机；短语介词"with the purpose of"表示目的；"where the real hardware is not available for use"是状语从句；动名词短语"providing a platform to run programs…"和"having multiple instances…"都作 of 的介词宾语。

4. An essential characteristic of a Virtual Machine is that the software running inside is limited to the resources and other abstractions provided by the Virtual Machine—it cannot break out of its virtual environment.

译文：虚拟机的一个基本特征就是，内部运行的软件仅限于由虚拟机提供的资源和其他抽象的东西，因为它无法突破其虚拟环境。

分析：句中"that the software running inside is limited to… the Virtual Machine"是表语从句，在表语从句中的过去分词短语"provided by the Virtual Machine"作定语，修饰、限定其前的名词"the resources and other abstractions"；破折号后的内容是补充说明。

Module 3　Consolidation Exercise 巩固练习

K12.PDF

I. Answer the following questions according to the passage

1. What is the term Network Operating System used to refer to?

2. What essential characteristic of a Virtual Machine is mentioned in Text 1?

3. What does a system Virtual Machine provide?

4. What is the replacement of NetWare 6.5SP8 General Support?

5. Is Solaris a Unix operating system? Who originally developed it?

6. Can Linux be installed on any computer hardware?

II. Fill in each blank with information from the passage

1. The term Network Operating System is used to refer to _____ for a network device such as a _____, _____ or _____ that operates the functions in the network layer (layer 3).

2. Like Windows Vista and _____, Windows Server 2008 is built on Windows 6.x. Therefore, it shares much of the same _____ and _____.

3. Microsoft has announced that _____ is the last _____ Windows server operating system. It is available in some _____, similar to Windows Server 2003.

4. An essential characteristic of a Virtual Machine is that the software running inside is limited to the_____and _____ provided by the Virtual Machine—it cannot _____ of its virtual environment.

5. Since _____ is common, Windows Server 2008 Windows Server 2008 _____ comes with most of the _____, _____, _____ and administrative features new to _____ such as the _____(native IPv6, native wireless, speed and security improvements).

III. Translate the following sentences into Chinese

1. The term Network Operating System is also used to refer to an operating system oriented to computer networking, which allows shared file and printer access among multiple computers in a network.

2. Linux can be installed on a wide variety of computer hardware, ranging from mobile phones, tablet computers and video game consoles, to mainframes and supercomputers.

3. Linux is also a top contender in the desktop OS market, due to its extremely secure and stable nature, its speed, and its lack of fragmentation issues.

4. Because Linux is freely redistributable, it is possible for anyone to create a distribution for any intended use.

5. Solaris supports SPARC-based and x86-based workstations and servers from Oracle and other vendors, with efforts underway to port to additional platforms.

6. A Virtual Machine (VM) is a software implementation of a machine (for example, a computer) that executes programs like a physical machine.

7. A Process Virtual Machine (also, Language Virtual Machine) is designed to run a single program, which means that it supports a single process.

Unit Three　Network Principles and Network Architecture
网络原理与体系结构

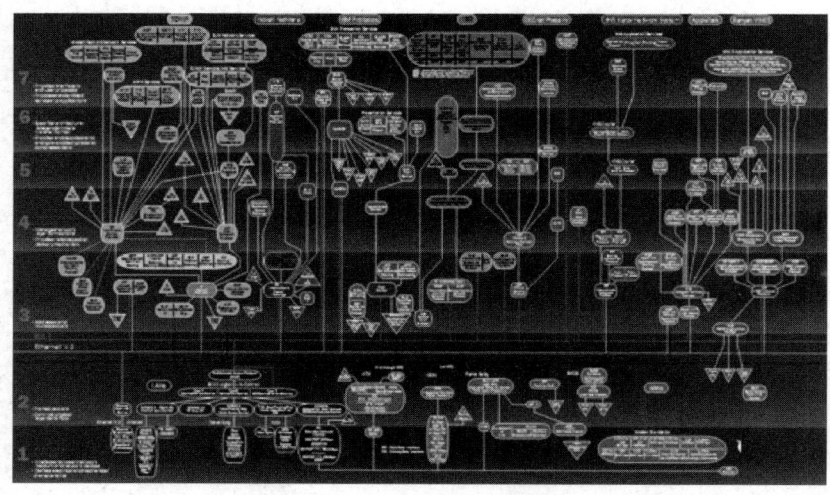

Lesson 13　Digital Transmission Technology

Module 1　Text Study 课文学习

Basic Training 基本训练

Text 1　根据语音和视频完成以下任务。

Task 1-0 听录音，记录关键词，理解课文大意。

L13-1-1.MP3

L13-1-2.MP3

_____

_____

_____

_____

_____

Task 1-1 词汇辨析。

① electric、electrical 和 electronic。

L13-1.MP4
_____

② digit、digital、numeral、figure 和 number。

_____

_____

③ convert、change、transform 和 turn。

_____

_____

Task 1-2 以定义或举例的方式，用英语解释以下术语。

① communication channel。

_____

② electromagnetic signal。

_____

Task 1-3 翻译画线部分或填空。

① 数据_____和数字_____不是一个概念，数字只是众多数据承载形式中的一种，目前，数据传输_____一般都是指数字化了的传输_____，只有在和模拟传输_____相比较的时候才会加以区分。

② information 和 message 都可翻译成"信息"，但_____侧重于人理解了的内容，是不可数名词（和 news 一样），"一条信息"要表达为_____，"许多信息"要表达为_____。message 和 data 一样，侧重于信息的载体，是可数名词。"通信"侧重于_____，"通讯"侧重于_____。

Task 1-4 句子翻译。

The messages are either represented by a sequence of pulses by means of a line code (baseband transmission), or by a limited set of continuously varying wave forms (passband transmission), using a digital modulation method.

_____

_____

_____

Task 1-5 解释术语及英汉互译。

PAM_____   ASK_____

FSK_____   PSK_____

串行电缆_____   并行信号_____

连续的_____   离散的_____

线路编码_____   电磁信号_____

检波_____   数模转换_____

Task 1-6 填空。

在本文中，_____指的是数字载波信号（Digital Carrier Signal），而_____指的是模拟载波信号（Analog Carrier Signal），但并不是所有文献都这样定义。

Baseband and Passband

While analog communication is the transfer of analog information originally represented by physical continuously varying analog signals, digital (data) communication is the transfer of discrete messages, using a digital modulation method, over digital or analog, point-to-point or point-to-multipoint communication channels. The discrete messages are either represented by a sequence of physical pulses, or by a limited set of physical continuously varying wave forms. Examples of such channels are copper wires, optical fibers, wireless communication channels, storage media and computer buses. The data (digital information) is represented as an electromagnetic signal, such as an electrical voltage, radio wave, microwave, or an infrared signal. The physically transmitted signal may be one of the following:

A baseband signal (digital information-over-digital transmission). A sequence of electrical pulses or light pulses produced by means of a line coding scheme such as Manchester Coding. This is typically used in serial cables, Wired Local Area Networks such as Ethernet, and in optical fiber communication. It results in a Pulse Amplitude Modulated (PAM) signal, also known as a pulse train, using in analog information-over-digital transmission.

A passband signal (digital information-over-analog transmission, Figure 3-1). A modulated sine wave signal (digitized analog signal) representing a digital bit-stream. The signal is produced by means of a digital modulation method such as PSK (Phase Shift Keying), ASK (Amplitude Shift Keying) or FSK (Frequency Shift Keying). The passband modulation and corresponding demodulation (also known as detection) are carried out by modem equipment. This is used in wireless communication, and over telephone network local-loop and cable-TV networks.

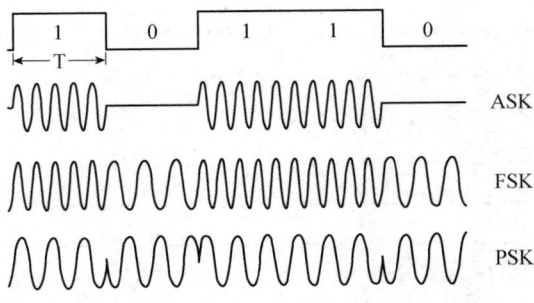

Figure 3-1   Passband Signal

Note: This classification is based on what is transmitted. According to the most common definition of digital communication, both baseband and passband signals representing bit-streams are considered as digital transmission, while an alternative definition only considers the digital baseband signal as digital communication, and passband transmission of digital data as a form of digital-to-analog conversion.

Data transmitted may be digital messages originating from a data source, for example, a computer or a keyboard. It may also be an analog signal such as a phone call or a video signal, digitized into a bit-stream for example using Pulse Code Modulation (PCM) or more advanced source coding (analog-to-digital conversion and data compression) schemes. This source coding and decoding is carried out by codec equipment.

Text 2　根据语音和视频完成以下任务。

Task 2-0 听录音，记录关键词，理解课文大意。

_____

_____

_____

L13-2.MP3

L13-2.MP4

Task 2-1 用英文填空。

① _____导致了背景噪声和信号失真。

② AXE 是一种用于电话交换机上的_____。

Task 2-2 术语解释及英汉互译。

ADSL_____　　ISDN_____

FTTB_____　　FTTH_____

视频会议_____　　数字无线电_____

多路技术_____　　时分多路复用_____

cable modem_____　　采样、抽样_____

Task 2-3 分析 telemetry 的构词。

_____

Task 2-4 用英文简述数字化传输模拟信号的优点。

_____

_____

_____

课文

Transmitting Analog Signals Digitally

In telephone networks, digital communication is utilized for transferring many phone calls over the same copper cable or fiber cable by means of Pulse Code Modulation (PCM), i.e., sampling and digitization, in combination with Time Division Multiplexing (TDM). Telephone exchanges have become digital and software has been controlled, facilitating many value added services. For example, the first AXE telephone exchange was presented in 1976. Since the late 1980s, digital communication to the end user has been possible using Integrated Services Digital Network (ISDN) services. Since the end of the 1990s, broadband access techniques such as ADSL, Cable modems, Fiber-To-The-Building (FTTB) and Fiber-To-The-Home (FTTH) have become widespread to small offices and homes. The current tendency is to replace traditional telecommunication services by packet mode communication such as IP telephony and IPTV.

Transmitting analog signals digitally allows for greater signal processing capability. The ability to process a communication signal means that errors caused by random processes can be detected and corrected. Digital signals can also be sampled instead of continuously monitored. The multiplexing of multiple digital signals is much simpler than the multiplexing of analog signals.

The digital revolution has also resulted in many digital telecommunication applications where the principles of data transmission are applied. Examples are second-generation and later cellular telephony, video conferencing, digital TV, digital radio, telemetry, etc.

Advanced Training 进阶训练

T13.PDF

Task 根据课文翻译下列专业术语。

无线对讲机＿＿＿＿＿＿＿＿＿＿＿＿＿＿＿   双工＿＿＿＿＿＿＿＿＿＿＿＿＿＿＿＿＿
单工＿＿＿＿＿＿＿＿＿＿＿＿＿＿＿＿＿＿＿   半双工＿＿＿＿＿＿＿＿＿＿＿＿＿＿＿＿
全双工＿＿＿＿＿＿＿＿＿＿＿＿＿＿＿＿＿＿   校验位＿＿＿＿＿＿＿＿＿＿＿＿＿＿＿＿
单播＿＿＿＿＿＿＿＿＿＿＿＿＿＿＿＿＿＿＿   广播＿＿＿＿＿＿＿＿＿＿＿＿＿＿＿＿＿
组播＿＿＿＿＿＿＿＿＿＿＿＿＿＿＿＿＿＿＿

Types of Communication Channels

A duplex communication system is a point-to-point system composed of two connected parties or devices that can communicate with one another in both directions. "Duplex" comes from "duo" that means "two", and "plex" that means "weave" or "fold"; thus, a duplex system has two clearly defined paths, with each path carrying information in only one direction: A to B over one path, and B to A over the other. There are two types of duplex communication systems: full-duplex and half-duplex. In a full-duplex system, both parties can communicate with each other simultaneously. An example of a full-duplex device is a telephone. In a half-duplex system, there are still two clearly defined paths/channels, and each party can communicate with the other but not simultaneously. An example of a half-duplex device is a walkie-talkie two-way radio that has a "push-to-talk" button.

Asynchronous and Synchronous Data Transmission

Asynchronous start-stop transmission uses start and stop bits to signify the beginning bit ，then ASCII character would actually be transmitted using 10 bits. For example, "01000001" would become "1010000010". The extra one (or zero, depending on parity bit) at the start and end of the transmission tells the receiver first that a character is coming and secondly that the character has ended. This method of transmission is used when data are sent intermittently as opposed to in a solid stream（连续的数据流）. In the previous example the start and stop bits are in bold. The start and stop bits must be of opposite polarity. This allows the receiver to recognize when the second packet of information is being sent.

Synchronous transmission uses no start and stop bits, but instead synchronizes transmission speeds at both the receiving and sending ends of the transmission using clock signal(s) built into each component. A continual stream of data is then sent between the two nodes. Due to there being no start and stop bits, the data transfer rate is quicker although more errors will occur, as the clocks will eventually get out of sync, and the receiving device would have the wrong time that had been agreed in the protocol for sending/receiving data, so some bytes could become corrupted (by losing bits). Ways to get around this problem include re-synchronization of the clocks and use of check digits to ensure the byte is correctly interpreted and received.

Transmission Technology

Broadly speaking, there are two types of transmission technology that are in widespread use:

Broadcast links and point-to-point links.

Point-to-point links connect individual pairs of machines. To go from the source to the destination on a network made up of point-to-point links, short messages, called packets in certain contexts, may have to first visit one or more intermediate machines. Often multiple routes, of different lengths, are possible, so finding good ones is important in point-to-point networks. Point-to-point transmission with exactly one sender and exactly one receiver is sometimes called unicasting.

In contrast, on a broadcast network, the communication channel is shared by all the machines on the network; packets sent by any machine are received by all the others. An address field within each packet specifies the intended recipient. Upon receiving a packet, a machine checks the address field. If the packet is intended for the receiving machine, that machine processes the packet; if the packet is intended for some other machine, it is just ignored.

Vocabulary Practice 词汇练习

词汇及短语听写，并纠正发音。

C13.MP3

Words & Expressions

discrete [dɪˈskriːt]　　　　　　 *adj.* 分离的，不相关的
codec [ˈkoʊdek]　　　　　　 *n.* 编解码器
multiplex [ˈmʌltɪpleks]　　　　 *vt.* 多路复用；多路传输　 *n.* 多厅影院
conference [ˈkɒnfərəns]　　　　 *n.* 会议，研讨会；讨论会
telemetry [təˈlemətrɪ]　　　　　 *n.* 遥感勘测，自动测量记录传导；测远术
amplitude [ˈæmplɪtjuːd]　　　　 *n.* 振幅
walkie-talkie [ˌwɔːkɪˈtɔːki]　　 *n.* 对讲机
asynchronous [eɪˈsɪŋkrənəs]　 *adj.* 异步的；不同时存在（或发生）的；非共时的
intermittently [ˌɪntəˈmɪtəntlɪ]　 *adv.* 间歇地；断断续续地
polarity [pəˈlærətɪ]　　　　　　 *n.* 极性；反向性；对立

Module 2　Study Aid 辅助帮学

Terms & Abbreviations 术语和缩写

baseband transmission　　　基带传输，按照信号原有的波形在信道上直接传输，要求信道具有较宽的频带。基带传输不需要调制、解调，设备花费少，适用于较小范围的信号传输。

passband transmission　　　通带传输，是一种采用调制、解调技术的传输形式。在发送端采用调制手段，对原始信号进行某种变换，例如，将代表数据的二进制"1"和"0"变换成具有一定频带范围的模拟信号，以适应在模拟信道上传输。

TDM　　 *abbr.* 时分复用模式（Time Division Multiplexing），是同时在同一个通信媒体上传输多个数字化数据、语音和视频信号的技术，其方法是把使用信道的时间分成多个片段

（时隙），不同的信号依次轮流占用不同时隙。

ISDN　　*abbr.* 综合业务数字网（Integrated Services Digital Network），是一个数字电话网络的国际标准，是一种典型的电路交换网络系统。

ADSL　　*abbr.* 非对称数字用户线路（Asymmetric Digital Subscriber Line），是一种新的数据传输方式，其上行和下行带宽不对称。

cable modem　　电缆调制解调器（也叫线缆调制解调器），是近几年随着网络应用的扩大而发展起来的，主要用于在有线电视网上进行数据传输。

FTTB　　*abbr.* 光纤到楼（Fiber-To-The-Building），是指光纤接到楼宇，通过双绞线接到各个用户。

FTTH　　*abbr.* 光纤到户（Fiber-To-The-Home），是指将光网络接入用户房间。根据光纤接近用户的程度，还有 FTTD（desk，桌面）、FTTC（curb，路边）、FTTZ（zone，小区）、FTTO（office，办公室）、FTTF（floor，楼层）、FTTSA（Service Area，服务区）等。

Manchester Coding　　曼彻斯特编码，也叫作相位编码（Phase Encode，PE），是一个同步时钟编码技术，被物理层用来编码一个同步位流的时钟和数据，常用于局域网传输。

PAM　　*abbr.* 脉冲振幅调制（Pulse Amplitude Modulated），是载波信号的振幅随基带信号变化的一种调制方式。

PSK　　*abbr.* 移相键控（Phase Shift Keying），用数字数据调制载波信号的相位。

QAM　　*abbr.* 正交振幅调制（Quadrature Amplitude Modulation），其幅度和相位同时变化，是正交载波调制技术与多电平振幅键控的结合。

FSK　　*abbr.* 频移键控（Frequency Shift Keying），是信息传输中使用得较早的一种调制方式，它的主要优点是实现起来较容易，抗噪声与抗衰减的性能较好。

local-loop　　本地回路，是连接电话公司与用户电话的有线连接，通常使用一对铜线连接。

cable-TV　　有线电视，使用同轴电缆传输电视信号。

duplex　　双工（Duplex Separation），指两台通信设备之间允许有双向的数据传输。数据通信中，数据在线路上的传送方式可以分为单工（simplex）通信、半双工（half-duplex）通信和全双工（full-duplex）通信三种。

parity bit　　奇偶校验位，是一个表示二进制数字串中"1"的个数是奇数还是偶数的二进制数，是最简单的错误校验码。

unicast　　单播，是一个发送者和一个接收者通过网络进行通信的传播方式。另外两种传播方式是组播（multicast）和广播（broadcast）。

Difficult Sentences Analysis and Translation 难句分析与翻译

1. The messages are either represented by a sequence of pulses by means of a line code (baseband transmission), or by a limited set of continuously varying wave forms (passband transmission), using a digital modulation method.

译文：信息通过一系列线性编码的脉冲（基带传输）表示，或者通过一组有限的连续变化波形（通带传输）表示，该组变化波形采用了数字调制方式。

分析：本句主干部分是"The messages are either represented by…, or by…"；"either…or…（或者……或者……）"前后短语是并列关系；用逗号隔开的现在分词短语"using a digital modulation method" 是方式状语，表示动作或状态发生的方式。

2. According to the most common definition of digital communication, both baseband and passband signals representing bit-streams are considered as digital transmission, while an alternative definition only considers the baseband signal as digital communication, and passband transmission of digital data as a form of digital-to-analog conversion.

译文：根据数字通信最常见的定义，代表比特流的基带和通带信号都被认为是数字传输；而另一种定义则只认为基带信号是数字通信，通带传输数字数据则是一种数字模拟转换形式。

分析：现在分词短语 "representing bit-streams" 作定语，修饰、限定前面的名词性短语 "both baseband and passband signals"；"while" 表示对比，有转折意义，句中动词短语 "consider…as" 意思是 "把……当作……；认为……是……"，并列连词 "and" 后省略了动词 "considers"。

3. Due to there being no start and stop bits, the data transfer rate is quicker although more errors will occur, as the clocks will eventually get out of sync, and the receiving device would have the wrong time that had been agreed in the protocol for sending/receiving data, so some bytes could become corrupted (by losing bits).

译文：虽然会由于没有起始和停止位出现更多的错误，但是数据传输的速率会更快。出现更多错误的原因是，时钟最后会失去同步，接收装置先前商定的发送/接收数据的协议时间会出错，所以有些字节就有可能因丢失数据位而损坏。

分析："Due to" 是短语介词，意思是 "因为，由于"；"there being no start and stop bits" 是动名词短语作介词宾语，"Due to there being no start and stop bits" 的意思是 "由于没有起始和停止位"；"although" 引导让步状语从句；在 "as" 引导的原因状语从句中，"the clocks will eventually get out of sync" 和 "the receiving device would have the wrong time" 是并列句，后句是虚拟语气，表示可能发生的结果；"that had been agreed in the protocol for sending/receiving data" 是定语从句，修饰、限定前面的 "time"；"so some bytes could become corrupted (by losing bits)" 也是虚拟语气，表示由此带来的 "字节损坏" 的可能性。

Module 3　Consolidation Exercise 巩固练习

K13.PDF

I. Give the meaning of the following abbreviated terms both in English and Chinese

PCM	TDM	ADSL	FTTB
PAM	PSK	ISDN	FSK

II. Translate the following sentences into Chinese

1. Data transmission is the physical transfer of data (a digital bit stream or a digitized analog signal) over a point-to-point or point-to-multipoint communication channels.

2. While analog transmission is the transfer of a continuously varying analog signal over an analog channel, digital communications is the transfer of discrete messages over a digital or an analog

channel.

3. In telephone networks, digital communication is utilized for transferring many phone calls over the same copper cable or fiber cable by means of Pulse Code Modulation (PCM), i.e., sampling and digitization, in combination with Time Division Multiplexing (TDM).

4. A duplex communication system is a point-to-point system composed of two connected parties or devices that can communicate with one another in both directions.

5. In a half-duplex system, there are still two clearly defined paths/channels, and each party can communicate with the other but not simultaneously.

6. The extra one (or zero, depending on parity bit) at the start and end of the transmission tells the receiver first that a character is coming and secondly that the character has ended.

7. To go from the source to the destination on a network made up of point-to-point links, short messages, called packets in certain contexts, may have to first visit one or more intermediate machines.

8. Upon receiving a packet, a machine checks the address field. If the packet is intended for the receiving machine, that machine processes the packet; if the packet is intended for some other machine, it is just ignored.

III. Make sentences with the following words

1. discrete

2. amplitude

3. asynchronous

4. intermittently

5. codec

Lesson 14   Communications Protocols

Module 1   Text Study 课文学习

Basic Training 基本训练

Text 1   根据语音和视频完成以下任务。

Task 1-0 听录音，记录关键词，理解课文大意。

L14-1-1.MP3

_____

_____

_____

L14-1-2.MP3

_____

_____

Task 1-1 翻译。

L14-1.MP4

错误恢复_____   物理量_____

实体_____   合作_____

结合_____   准确的_____

特别的_____   具体的_____

Task 1-2 填空。

① There is a close_____between protocols and programming languages. Protocols _____ to communications _____ programming languages _____ to computations.

② Communications protocols have to be _____ by the parties involved. To _____ agreement, a protocol may be _____ into a technical standard.

Task 1-3 词汇及词组辨析。

① function 和 functionality 都有"功能"的意思，但_____偏抽象，指"具有……的性质""有……设计目的"的含义。

② 可翻译为"基础"的词汇有 base 和 basis，_____较为抽象，而_____较为具体。

③ kind of、sort of 作为"种类"时，_____比_____更准确；_____是界限明确的分类，而_____有"级别"的含义。

④ for example 和 for instance 有什么区别？

_____

Task 1-4 翻译及词汇辨析。

① The data received has to be evaluated in the context of the progress of the conversation.

比较 process 和 progress 的区别。

_____

_____

② Other rules determine whether the data is meaningful for the context in which the exchange takes place.

比较 determine 和 decide 的区别。

_____

_____

③ There is a close analogy between protocols and programming languages.

…given the similarities between programming languages and communications protocols.

比较 analogy 和 similarity 的区别。

_____

_____

_____

④ Each message has an exact meaning intended to elicit a response from a range of possible responses pre-determined for that particular situation.

The specified behavior is typically independent of how it is to be implemented.

Each layer solves a distinct class of problems.

比较 exact、particular、distinct、typically 和 specific/specified 的区别。

_____

_____

_____

_____

课文

Protocols in Communicating Systems

In telecommunications, a communications protocol is a system of rules that allow two or more entities of a communications system to transmit information via any kind of variation of a physical quantity. These are the rules or a standard that defines the syntax, semantics and synchronization of communication and possible error recovery methods. Protocols may be implemented by hardware, software, or a combination of both.

Communicating systems use protocols, such as well-defined data formats, for exchanging messages. Each message has an exact meaning intended to elicit a response from a range of possible responses pre-determined for that particular situation. The specified behavior is typically independent of how it is to be implemented. Communications protocols have to be agreed upon by the parties involved. To reach agreement, a protocol may be developed into a technical standard. A programming language describes the same for computations, so there is a close analogy between protocols and programming languages. Protocols are to communications as programming languages are to computations.

To implement a networking protocol, the protocol software modules are interfaced with a framework implemented on the machine's operating system. This framework implements the networking functionality of the operating system. The best-known frameworks are the TCP/IP model and the OSI model.

At the time the Internet was developed, layering had proven to be a successful design approach for both compiler and operating system design and, given the similarities between programming languages and communications protocols, layering was applied to the protocols as well. This gave rise to the concept of layered protocols which nowadays forms the basis of protocol design.

Systems typically do not use a single protocol to handle a transmission. Instead, they use a set of cooperating protocols, sometimes called a protocol family or a protocol suite. Some of the best-known protocol suites include: IPX/SPX, X.25, AX.25, AppleTalk and TCP/IP.

The protocols can be arranged based on functionality in groups, for instance, there is a group of transport protocols. The functionalities are mapped onto the layers, each layer solving a distinct class of problems relating to, for instance, application-, transport-, internet- and network interface-functions. To transmit a message, a protocol has to be selected from each layer, so some sort of multiplexing/demultiplexing takes place. The selection of the next protocol is accomplished by extending the message with a protocol selector for each layer.

Getting the data across a network is only part of the problem for a protocol. The data received has to be evaluated in the context of the progress of the conversation, so a protocol has to specify rules describing the context. These kinds of rules are said to express the syntax of the communications. Other rules determine whether the data is meaningful for the context in which the exchange takes place. These kinds of rules are said to express the semantics of the communications.

Text 2   根据语音和视频完成以下任务。

Task 2-0 听录音，记录关键词，理解课文大意。

L14-2-1.MP3

L14-2-2.MP3

L14-2-3.MP3

Task 2-1 英汉互译及解释。

① MTU_____

② CRC_____

③ MAC_____

L14-2.MP4

④ 寻址方案_____

⑤ 地址映射_____

⑥ 面向连接的通信_____

⑦ 超时_____

⑧ 这是一个全部是"1"的地址_____

⑨ Error-free_____

⑩ Broken link_____

Task 2-2 翻译。

① 超过最大传输单元的长比特流被分成块，每个块是一个数据包，每个数据包被分成两个部分，即头部区域和负载区域，每个区域又分成不同的字段。

_____

_____

_____

② The pieces may get lost or delayed or take different routes to their destination on some types of networks.

_____

③ Pieces may arrive out of sequence.

_____

Task 2-3 比较 intermediary 和 intermediate 的词性和词义。

_____

Task 2-4 填空。

① Protocols should therefore specify rules _____ the transmission. In _____, much of the following should be_____.

② The receiver _____ the packets _____ CRC differences and _____ somehow for retransmission.

③ Packets may be _____ on the network or _____ from long delays.

④ Flow control can be implemented by _____ from _____ to _____.

课文

Basic Requirements of Protocols

Messages are sent and received on communicating systems to establish communications. Protocols should therefore specify rules governing the transmission. In general, much of the following should be addressed:

● Data Formats for Data Exchange.

The exchanged digital message bit strings are divided in fields and each field carries information relevant to the protocol. Conceptually the bit string is divided into two parts called the header area and the data area. The actual message is stored in the data area, so the header area contains the fields with more relevance to the protocol. Bit strings longer than the Maximum Transmission Unit (MTU) are divided in pieces of appropriate size.

Addresses are used to identify both the sender and the intended receiver(s). The addresses are stored in the header area of the bit strings, allowing the receivers to determine whether the bit strings are intended for themselves and should be processed or should be ignored. A connection between a sender and a receiver can be identified using an address pair (sender address, receiver address). Usually, some address values have special meanings. An all 1s address could be taken to mean an addressing of all stations on the network, so sending to this address would result in a broadcast on the local network. The rules describing the meanings of the address value are collectively called an addressing scheme.

- Address Mapping.

Sometimes protocols need to map addresses of one scheme on addresses of another scheme, for instance, to translate a logical IP address specified by the application to an Ethernet hardware address. This is referred to as address mapping.

- Routing.

When systems are not directly connected, intermediary systems along the route to the intended receiver(s) are needed to forward messages on behalf of the sender. On the Internet, the networks are connected using routers. This way of connecting networks is called internetworking.

- CRC (Cyclic Redundancy Check).

Detection of transmission errors is necessary on networks which cannot guarantee error-free operation. In a common approach, CRCs of the data area are added to the end of packets, making it possible for the receiver to detect differences caused by errors. The receiver rejects the packets on CRC differences and arranges somehow for retransmission.

- Acknowledgements.

Acknowledgements of correct reception of packets are required for connection-oriented communication. Acknowledgements are sent from receivers back to their respective senders.

- Loss of Information—Timeouts and Retries.

Packets may be lost on the network or suffer from long delays. To cope with this, under some protocols, a sender may expect an acknowledgement of correct reception from the receiver within a certain amount of time. On timeouts, the sender must assume the packet was not received and retransmit it. In case of a permanently broken link, the retransmission has no effect so the number of retransmissions is limited. Exceeding the retry limit is considered an error.

- Access Control.

Direction of information flow needs to be addressed if transmissions can only occur in one direction at a time as on half-duplex links. This is known as Media Access Control (MAC). Arrangements have to be made to accommodate the case when two parties want to gain control at the same time.

- Sequence Controlling.

We have seen that long bit strings are divided in pieces, and then sent on the network individually. The pieces may get lost or delayed or take different routes to their destination on some types of networks. As a result，pieces may arrive out of sequence. Retransmissions can result in duplicate pieces. Through marks on the pieces with sequence information by the sender, the receiver can determine what was lost or duplicated, ask for necessary retransmissions and reassemble the original message.

- Flow Control.

Flow control is needed when the sender transmits faster than the receiver or when intermediate network equipment cannot process the transmissions. Flow control can be implemented by messaging from receiver to sender.

Advanced Training 进阶训练

T14.PDF

Task 根据课文翻译并解释术语。

① ISO_____

② ITU_____

③ IEEE_____

④ IETF_____

⑤ W3C_____

Communication-Related Standards Organizations

Some of the standards organizations of relevance for communications protocols are the International Organization for Standardization (ISO) (Figure 3-2), the International Telecommunication Union (ITU), the Institute of Electrical and Electronics Engineers (IEEE), and the Internet Engineering Task Force (IETF). The IETF maintains the protocols in use on the Internet.

Figure 3-2   ISO Sign

The IEEE controls many software and hardware protocols in the electronics industry for commercial and consumer devices. The ITU is an umbrella organization of telecommunication engineers designing the Public Switched Telephone Network (PSTN), as well as many radio communications systems. For marine electronics the NMEA standards are used. The World Wide Web Consortium (W3C) produces protocols and standards for Web technologies.

International standards organizations are supposed to be more impartial than local organizations with a national or commercial self-interest to consider. Standards organizations also do research and development for standards of the future. In practice, the standards organizations mentioned, cooperate closely with each other.

Vocabulary Practice 词汇练习

C14.MP3

词汇及短语听写，并纠正发音。

_____

_____

_____

Words & Expressions

syntax ['sɪntæks]　　　　　　*n.* 句法；语构；语法

semantics [sɪ'mæntɪks]　　　*n.* 语义学；词义学

elicit [ɪ'lɪsɪt]　　　　　　　*vt.* 引出；探出；诱出

analogy [əˈnælədʒɪ]	*n.* 类似，相似；比拟，类比；类推
compiler [kəmˈpaɪlə(r)]	*n.* 汇编者；编辑者；编译程序
demultiplex [dɪˈmʌltɪpleks]	*vt.* [电信]解多路复用；信号分离
intermediary [ˌɪntəˈmiːdɪərɪ]	*adj.* 中间人的；调解的；居间的；媒介的 *n.* 调解人
acknowledgement [əkˈnɒlɪdʒmənt]	*n.* 确认，承认；感谢；谢礼
timeout [taɪmˈaʊt]	*n.* 暂时休息，（工作时的）工间休息
duplicate [ˈdjuːplɪkeɪt]	*adj.* 复制的，复写的；完全一样的
	vt. 复制；重复
	n. 复制品；副本
context [ˈkɒntekst]	*n.* 上下文；背景；环境；语境
impartial [ɪmˈpɑːʃl]	*adj.* 不偏不倚的；公平的；持平

Module 2　Study Aid 辅助帮学

Terms & Abbreviations 术语和缩写

SPX *abbr.* 顺序分组交换（Sequences Packet Exchange），是 Novell 公司的通信协议集。

X.25 X.25 协议，一个广泛使用的网络协议，由 ITU-T 提出，是面向计算机的数据通信网。

AX.25 AX.25 协议是为业余无线分组网络通信设计的一个数据链接层协议，由 X.25 协议衍生而来，占据了 OSI 网络模型的第一和第二层，支持有连接和无连接两种操作模式，后者常被用于自动位置报告系统中。

AppleTalk 苹果交流协议，是由苹果公司创建的一组网络协议，用于苹果系列的个人计算机。协议栈中的各种协议用来提供通信服务，例如，文件服务、打印、电子邮件和一些其他网络服务。

MTU *abbr.* 最大传输单元（Maximum Transmission Unit），是指一种通信协议的某个层上面所能通过的最大数据包大小。最大传输单元这个参数通常与通信接口（网络接口卡、串口等）有关。

CRC *abbr.* 循环冗余校验（Cyclic Redundancy Check），是数据通信领域中最常用的一种差错校验码。它的特征是信息字段和校验字段的长度可以任意选定，对数据进行多项式计算，并将得到的结果附在帧的后面，接收设备也执行类似的算法，以保证数据传输的正确性和完整性。

ITU *abbr.* 国际电信联盟（International Telecommunication Union），简称国际电联，是主管信息通信技术事务的联合国机构，负责分配和管理全球无线电频谱与卫星轨道资源，制定全球电信标准，向发展中国家提供电信援助，促进全球电信发展。

NMEA *abbr.* 美国国家海洋电子协会（National Marine Electronics Association），为海用电子设备制定标准格式。

Difficult Sentences Analysis and Translation 难句分析与翻译

1. In telecommunications, a communications protocol is a system of rules that allow two or more

entities of a communications system to transmit information via any kind of variation of a physical quantity.

译文：在电信中，通信协议是一种允许通信系统的两个或多个实体通过物理量的任何变化来传输信息的规则系统。

分析：本句中"that"引导定语从句，修饰、限定前面的"rules"，"that"在从句中作主语；介词短语"via any kind of variation of a physical quantity"的意思是"通过物理量的任何种类的变化"。

2. At the time the Internet was developed, layering had proven to be a successful design approach for both compiler and operating system design and, given the similarities between programming languages and communications protocols, layering was applied to the protocols as well.

译文：在互联网发展的时候，对编译器和操作系统设计来说，分层已经被证明是一种成功的设计方法。考虑到编程语言和通信协议之间的相似之处，分层也应用于协议。

分析：本句包含"At the time the Internet was developed, layering had proven to be a…operating system design and"和"given the similarities…layering was applied to the protocols as well"两个并列的句子；前句中的"At the time the Internet was developed"作时间状语，后半句"given the similarities between programming languages and communications protocols"中的"given"是介词，意思是"考虑到、鉴于"。

3. The functionalities are mapped onto the layers, each layer solving a distinct class of problems relating to, for instance, application-, transport-, internet- and network interface-functions.

译文：这些功能被映射到各个层上，每层都解决不同类别的相关问题，例如，应用程序问题、传输问题、互联网问题及网络接口功能问题。

分析：本句中的"each layer solving a distinct class of problems relating to, for instance, application-, transport-, internet- and network interface-functions"是独立主格结构，在翻译时可以独立成句。

4. The addresses are stored in the header area of the bit strings, allowing the receivers to determine whether the bit strings are intended for themselves and should be processed or should be ignored.

译文：这些地址存储在位串的头部区域中，允许接收器确定位串是否为自己使用、是否应该处理或者应该忽略。

分析：本句中的现在分词短语"allowing the receivers to determine…"作状语，所表达的动作伴随着主句谓语动词的动作而发生；"whether"引导宾语从句。

5. Through marks on the pieces with sequence information by the sender, the receiver can determine what was lost or duplicated, ask for necessary retransmissions and reassemble the original message.

译文：通过发送方在信息片段上标记的序列号，接收者可以确定丢失或重复的内容，要求必要的重传，并重新组合原始消息。

分析：句中的介词短语"Through marks on…by the sender"作方式状语；"what was lost or duplicated"是宾语从句，作动词"determine"的宾语；动词"determine""ask for""reassemble"是并列关系，共同的主语是"the receiver"。

Module 3　Consolidation Exercise 巩固练习

I. Answer the following questions according to the passage

1. What does a communications protocol in telecommunications refer to?

2. What is the analogy between protocols and programming languages?

3. What are the best-known protocol frameworks that implement the networking functionality of the operating system?

4. What bit strings are divided in pieces of appropriate size?

5. What does an all 1s address mean?

6. What is referred to as address mapping?

7. Why is the number of retransmissions limited?

8. When is flow control needed?

9. List some standards organizations that are associated with communication protocols.

II. Fill in each blank with information from the passage

1. In telecommunications, a communications protocol is _____ that allow two or more _____ of a communications system to _____ via any kind of variation of a _____.

2. Communicating systems use _____, such as well-defined data formats, for _____. Each message has an exact meaning intended to _____ from a range of possible responses _____ for that particular situation.

3. Systems typically do not use a _____ to handle a transmission. Instead they use a set of _____ protocols, sometimes called a protocol family or a protocol _____.

4. Conceptually the bit string is divided into two parts called _____ and _____. The _____ is stored in the data area, so the header area contains the fields _____.

5. The addresses are stored in the header area of the bit strings, allowing the _____ to determine whether the bit strings are intended for themselves and should be _____ or should be _____.

6. Packets may be _____ on the network or _____ from long delays. To cope with this, under some protocols, a sender may expect an _____ of correct reception from the receiver within a certain amount of time. On _____, the sender must assume the packet was not received and _____ it.

7. We have seen that long bit strings are divided in _____, and then sent on the network _____. The pieces may _____ or _____ or _____ to their destination on some types of networks.

8. International standards organizations are supposed to be more _____ than local organizations with a national or commercial _____ to consider. Standards organizations also do _____ and development for standards of the future.

III. Translate the following sentences into English

1. 为了达成一致，协议可能会发展成为一项技术标准。

2. 比最大协议单元长的数据流将被分成适当大小的片段。

3. 有时协议需要将一种方案的地址映射为另一种方案的地址，例如：将 IP 地址映射为以太网硬件地址。

4. 通常的方法是在包的尾部增加循环冗余码，使接收方有可能检测到因错误造成的数据变化。

5. 确认信息由接收方发回各自的发送方。

6. 如果传输在同一时间只能在一个方向发生，比如在半双工链接上，则需要定位信息流的方向。

7. 通过发送方在信息片段上标记的序列号，接收方可以确定丢失或重复的内容，要求必要的重传，并重新组合原始消息。

8. 国际标准组织应该比需要考虑国家或商业利益的地方组织更为公正。

Lesson 15　CSMA/CD and IEEE 802

Module 1　Text Study 课文学习

Basic Training 基本训练

Text 1　根据语音和视频完成以下任务。

Task 1-0 听录音，记录关键词，理解课文大意。

L15-1-1.MP3

L15-1-2.MP3

L15-1.MP4

Task 1-1 英汉互译。

CSMA/CD_____

阻塞信号_____　时间间隔_____　32-bit binary pattern_____

来回时间_____　16 位 1 和 0 的组合_____　冲突域_____

64 octets_____　不匹配的_____　now-obsolete_____

IEEE Std 802.3_____　向后兼容_____　向前兼容_____

Task 1-2 翻译及词义辨析。

enhance_____　improve_____　increase_____

discard_____　ignore_____　cause_____

reason_____

Task 1-3 构词训练。

isolate 和 insulate 词义相近，是否具有相同的词源？

_____

课文

Carrier Sense Multiple Access with Collision Detection

Carrier Sense Multiple Access with Collision Detection (CSMA/CD) is a Media Access Control method used most notably in Local Area Network using early Ethernet technology. It uses a carrier sensing scheme in which a transmitting data station detects other signals while transmitting a frame, stops transmitting that frame, transmits a jam signal, and then waits for a random time interval before trying to resend the frame. CSMA/CD is a modification of pure Carrier Sense Multiple Access (CSMA). CSMA/CD is used to improve CSMA performance by terminating transmission as soon as a collision is detected, thus shortening the time required before a retry can be attempted.

The jam signal or jamming signal is a signal that carries a 32-bit binary pattern sent by a data station to inform the other stations of the collision that they must not transmit. The maximum jam-time is calculated as follows. The maximum allowed the diameter of an Ethernet installation is limited to 232 bits. This makes a round-trip-time of 464 bits. As the slot time in Ethernet is 512 bits, the difference between slot time and round-trip-time is 48 bits (6 bytes), which is the maximum "jam-time". This in turn means: A station noting a collision has occurred is sending a 4 bytes long pattern composed of 16 1-0 bits combinations. Note: The size of this jam signal is clearly beyond the minimum allowed frame-size of 64 bytes. The purpose of this is to ensure that any other node which may currently be receiving a frame will receive the jam signal in place of the correct 32-bit MAC CRC (Cyclic Redundancy Check), which causes the other receivers to discard the frame due to a CRC error.

In 10 megabit shared medium Ethernet, if a collision error occurs after the first 512 bits of data are transmitted by the transmitting station, a late collision is said to have occurred. Importantly, late collisions are not re-sent by the NIC unlike collisions occurring before the first 64 octets; it is left for the upper layers of the protocol stack to determine that there was loss of data. As a correctly set up CSMA/CD network link should not have late collisions, the usual possible causes are full-duplex/half-duplex mismatch, exceeded Ethernet cable length limits, or defective hardware such as incorrect cabling, non-compliant number of hubs in the network, or a bad NIC.

CSMA/CD was used in now-obsolete shared media Ethernet variants (10BASE5, 10BASE2) and in the early versions of twisted-pair Ethernet which used repeater hubs. Modern Ethernet networks, built with switches and full-duplex connections, no longer need to utilize CSMA/CD because each Ethernet segment, or collision domain, is now isolated. CSMA/CD is still supported for backwards compatibility and for half-duplex connections. IEEE Std 802.3, which defines all Ethernet variants, for historical reasons still bears the title "Carrier Sense Multiple Access with Collision Detection (CSMA/CD) access method and Physical Layer specifications".

Text 2　根据语音和视频完成以下任务。

Task 2-0 听录音，记录关键词，理解课文大意。

_____

_____

L15-2.MP3

_____

Task 2-1 填空。

shorten 和 liken 都是形容词加_____构成的动词。

L15-2.MP4

Task 2-2 替换练习。

① 分别用 analogy 和 similarity 替换 liken，表达：This can be likened to what happens at a dinner party.

_____

_____

② 分别用 base 和 relevant 替换 dependent，表达：Methods for collision detection are media dependent.

_____

_____

Task 2-3 翻译及理解。

stop_____ terminate_____ end_____

abort_____ finish_____ complete_____

initial_____ start_____ back off_____

identify_____ recognize_____ realize_____

make out_____ if yes_____ if so_____

Task 2-4 模仿课文，尝试用英语描述如下流程图（Figure 3-3）。

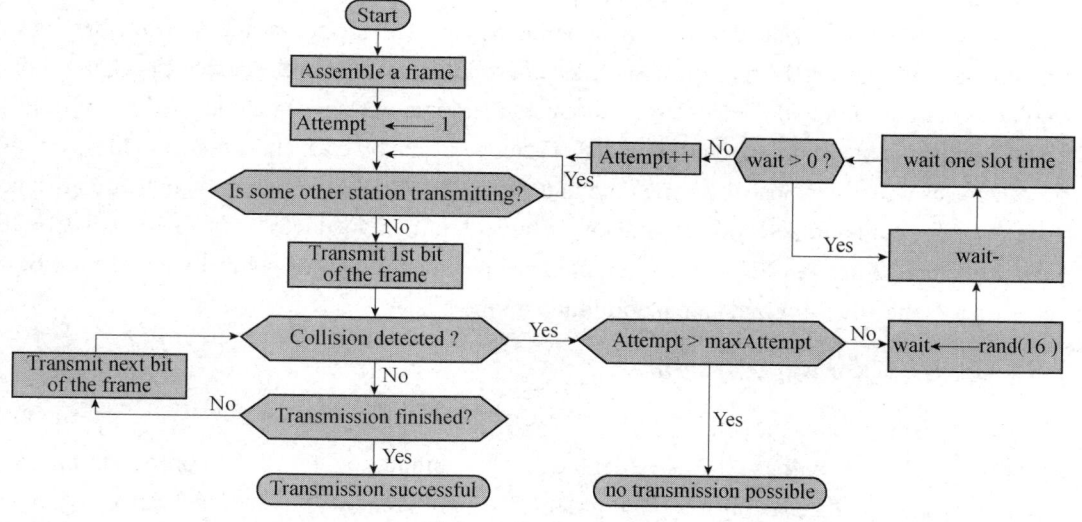

Figure 3-3   Flow Chart

_____

_____

_____

_____

_____

_____

_____

_____

课文

The Procedure of Transmission

The following procedure is used to initiate a transmission. The procedure is complete when the frame is transmitted successfully or a collision is detected during transmission.

- Is my frame ready for transmission? If yes, it goes on to the next point. Is medium idle? If not, wait until it becomes ready.
- Start transmitting and monitor for collision during transmission.
- Did a collision occur? If so, go to collision detected procedure.
- Reset retransmission counters and end frame transmission.

The following procedure is used to resolve a detected collision. The procedure is complete when retransmission is initiated or the retransmission is aborted due to numerous collisions.

- Continue transmission (with a jam signal instead of frame header/data/CRC) until minimum packet time is reached to ensure that all receivers detect the collision.
- Increase retransmission counter.
- Was the maximum number of transmission attempts reached? If so, abort transmission.
- Calculate and wait random back off period based on number of collisions.
- Re-enter main procedure from the start point.

This can be likened to what happens at a dinner party, where all the guests talk to each other through a common medium (the air). Before speaking, each guest politely waits for the current speaker to finish. If two guests start speaking at the same time, both stop and wait for short, random periods of time (in Ethernet, this time is measured in microseconds). The hope is that by each choosing a random period of time, both guests will not choose the same time to try to speak again, thus avoiding another collision.

Methods for collision detection are media dependent, but on an electrical bus such as 10BASE5 or 10BASE2, collisions can be detected by comparing transmitted data with received data or by recognizing a higher-than-normal signal amplitude on the bus.

Advanced Training 进阶训练

T15.PDF

Task 参考提示，将下列词汇填入下文。

steady	splits	denote	brand	contention-based
cell relay	generally	regular	reference	restricted
uniformly sized	revisions	subsequent	frequency bands	architecture
amendments	variable-size	by contrast	eliminating	revoked

IEEE 802

IEEE 802 refers to a family of IEEE standards dealing with Local Area Networks and Metropolitan Area Networks. More specifically, the IEEE 802 standards are _____（限制）to networks carrying _____（可变大小）packets. _____（相比之下）, in _____（信源中继）networks, data is transmitted in short, _____（统一大小）units called cells. Isochronous networks, where data is transmitted as a _____（恒定的）stream of octets, or groups of octets, at _____（定期的）time intervals, are also out of the scope of this standard. The number 802 was simply the next free number IEEE could assign, though "802" is sometimes associated with the date the first meeting was held — February 1980.

The services and protocols specified in IEEE 802 map to the lower two layers (Data Link and Physical) of the seven-layer OSI networking _____ （参考）model. In fact, IEEE 802 _____ （分解）the OSI Data Link Layer into two sub-layers named Logical Link Control (LLC) and Media Access Control (MAC), so that the layers can be listed like this:

- Data Link Layer.
  - ♦ LLC Sublayer.
  - ♦ MAC Sublayer.
- Physical Layer.

The IEEE 802 family of standards is maintained by the IEEE 802 LAN/MAN Standards Committee (LMSC). The most widely used standards are for the Ethernet family, Token Ring, Wireless LAN, Bridging and Virtual Bridged LANs. An individual Working Group provides the focus for each area.

IEEE 802.3 is a working group and a collection of IEEE standards produced by the working group defining the Physical Layer and Data Link Layer's Media Access Control (MAC) of wired Ethernet. This is _____ （通常）a Local Area Network technology with some Wide Area Network applications. Physical connections are made between nodes and/or infrastructure devices (hubs, switches, routers) by various types of copper or fiber cable. IEEE 802.3 is a technology that supports the IEEE 802.1 network _____ （架构）. IEEE 802.3 also defines LAN access method using CSMA/CD.

Introduced by IBM in 1984, it was then standardized with protocol IEEE 802.5 and was fairly successful, particularly in corporate environments, but gradually eclipsed by the later versions of Ethernet. Token Ring Local Area Network (LAN) technology is a communications protocol for Local Area Networks. It uses a special three-byte frame called a "token" that travels around a logical "ring" of workstations or servers. This token passing is a channel access method providing fair access for all stations, and _____ （消除）the collisions of _____ （基于竞争的）access methods.

IEEE 802.11 is a set of MAC and Physical Layer (PHY) specifications for implementing Wireless Local Area Network (WLAN) computer communication in the 900 MHz and 2.4, 3.6, 5, and 60 GHz _____ （频带）. They are created and maintained by the Institute of Electrical and Electronics Engineers (IEEE) LAN/MAN Standards Committee (IEEE 802). The base version of the standard was released in 1997, and has had _____ （后续的）amendments. The standard and _____ （修正）provide the basis for wireless network products using the Wi-Fi _____ （商标）. While each amendment is officially _____ （取消）when it is incorporated in the latest version of the standard, the corporate world tends to market to the _____ （校订）because they concisely _____ （表示）capabilities of their products. As a result, in the market place, each revision tends to become its own standard.

Vocabulary Practice 词汇练习

词汇及短语听写，并纠正发音。

_____ C15.MP3

_____

_____

Words & Expressions

notably [ˈnəʊtəblɪ]	*adv.* 尤其，特别
abort [əˈbɔːt]	*vt.* （使）中止；（使）夭折；（使）流产
increment [ˈɪŋkrəmənt]	*n.* 增量，增加，增长
liken [ˈlaɪkən]	*vt.* 比拟；把……比作
octet [ɒkˈtet]	*n.* 8 位位组，8 位字节
non-compliant [ˌnɒnkəmˈplaɪənt]	*adj.* 不符合规定的；非顺应性的；不顺从的
variant [ˈveərɪənt]	*n.* 变体；变种；变形　*adj.* 不同的；变体的；变种的
bear [beə(r)]	*vt.* 承担；忍受；支撑；生育
metropolitan [ˌmetrəˈpɒlɪtən]	*adj.* 大城市的，大都市的
isochronous [aɪˈsɒkrənəs]	*adj.* 同步的
eclipse [ɪˈklɪps]	*vt.* 使相形见绌，使黯然失色　*n.* 黯然失色，暗淡；日食；月食
amendment [əˈmendmənt]	*n.* 修改，修订；修正案
brand [brænd]	*n.* 商标，牌子
revoke [rɪˈvəʊk]	*vt.* 废除；撤销，取消
concisely [kənˈsaɪslɪ]	*adv.* 简明地；简洁地
denote [dɪˈnəʊt]	*vt.* 代表；指代；预示；意思是

Module 2　Study Aid 辅助帮学

Terms & Abbreviations 术语和缩写

10BASE2　　以太网标准，"10"表示传输速率为 10Mbps，"BASE"表示采用基带传输技术，"2"表示最大距离为 200m，被称为细缆网，在单个网段上最多可支持 30 个工作站。此外还有 10BASE5，该标准用于粗同轴电缆、速度为 10Mbps 的基带局域网络，在总线型网络中，最远传输距离为 500m。

jam signal　　阻塞信号，通过发送干扰信号让对方暂停发送信息，在解除阻塞后继续传递。

Std　*abbr.* 标准/样品（Standard），也可作形容词，意思是"标准的、一流的"。

cell　　信元，指长度固定、大小一致的数据包。信元中继（cell relay）是一个固定长度信息包的网络技术。

LLC　*abbr.* 逻辑链路控制（Logical Link Control），负责识别网络层协议，并对它们进行封装。

Token Ring　　令牌环，一种 LAN 协议，定义在 IEEE 802.5 中，其中所有的工作站都连接到一个环上，每个工作站只能向直接相邻的工作站传输数据。通过围绕环的令牌信息授予工作站传输权限。

PHY　*abbr.* 物理层（Physical Layer），OSI 模型的底层，可以指与信号接口的芯片。

Wi-Fi　　*abbr.* 无线保真（Wireless Fidelity），基于 IEEE 802.11b 标准的无线局域网。Wi-Fi 是一个无线网络通信技术的品牌，由 Wi-Fi 联盟所持有。

Difficult Sentences Analysis and Translation 难句分析与翻译

1. It uses a carrier sensing scheme in which a transmitting data station detects other signals while transmitting a frame, stops transmitting that frame, transmits a jam signal, and then waits for a random time interval before trying to resend the frame.

译文：它使用的是载波侦听方案，在该方案中，发送数据站在发送帧时，如果检测到其他信号就会停止发送该帧，发送一个阻塞信号，之后等待一段随机间隔时间再尝试重新发送该帧。

分析：本句中 "in which" 引导定语从句，"which" 指代前面的 "a carrier sensing scheme"；在定语从句中，"while transmitting a frame" 和 "before trying to resend the frame" 是时间状语从句，动词 "detects" "stops" "transmits" "waits" 是先后发生的系列动作。

2. CSMA/CD is used to improve CSMA performance by terminating transmission as soon as a collision is detected, thus shortening the time required before a retry can be attempted.

译文：使用 CSMA/CD，一旦检测到冲突就会通过终止传输来提高 CSMA 的性能，从而缩短重试之前所需的时间。

分析：本句中的 "CSMA/CD（带冲突检测的载波监听多路访问）" 和 "CSMA（载波监听多路访问）" 是术语缩写，在翻译时也可以不译；介词短语 "by terminating transmission" 表示方式；"as soon as a collision is detected" 和 "before a retry can be attempted." 是时间状语从句；现在分词短语 "thus shortening the time required…" 作状语，表示结果。

3. The hope is that by each choosing a random period of time, both guests will not choose the same time to try to speak again, thus avoiding another collision.

译文：目的是每个客户随机选择一个时间段，两个客户都不会再选择同一时间发送信息，从而可以避免另一次碰撞。

分析：本句中 "that" 引导表语从句；动名词短语 "choosing a random period of time" 有自己的逻辑主语 "each"，整个动名词复合结构作介词 "by" 的宾语，一起表示方式；现在分词短语 "thus avoiding another collision" 作状语，表示结果。

4. The purpose of this is to ensure that any other node which may currently be receiving a frame will receive the jam signal in place of the correct 32-bit MAC CRC, which causes the other receivers to discard the frame due to a CRC error.

译文：这样做的目的是确保当前正在接收帧的任何其他节点都能接收到代替正确的 32 位 MAC CRC 的阻塞信号，这就会导致其他接收者由于 CRC 错误而丢弃该帧。

分析：本句中的 "MAC（媒体访问控制）" 和 "CRC（循环冗余校验）" 是术语缩写，在翻译时也可以不译；动词不定式短语 "to ensure…" 作表语；"that any other node…will receive the jam signal…CRC" 从句作动词 "ensure" 的宾语，该宾语从句中的 "which may currently be receiving a frame" 是定语从句，修饰、限定 "any other node"；"which" 引导的非限制定语从句修饰整个主句，对主句所叙述的情况进行某种意义的补充说明，可译为 "这（一点），这样"。

5. Isochronous networks, where data is transmitted as a steady stream of octets, or groups of

octets, at regular time intervals, are also out of the scope of this standard.

译文：同步网络也不在本标准的范围之内，其数据作为稳定的 8 位字节流或 8 位位组以规则的时间间隔源源不断地传输。

分析：句中主干部分是"Isochronous networks are also out of the scope of this standard"；"where data is transmitted as a steady stream of octets, or groups of octets, at regular time intervals"是非限制性定语从句，翻译时可以单独成句。

6. While each amendment is officially revoked when it is incorporated in the latest version of the standard, the corporate world tends to market to the revisions because they concisely denote capabilities of their products.

译文：虽然每一次的修改在纳入最新版本的标准中都会被正式撤销，但企业界倾向于推广修订版，因为它们简明地表达了其产品的能力。

分析：句中的"While each amendment is officially revoked…"是让步状语从句，其中的"when it is incorporated in the latest version of the standard"是时间状语从句；"because they concisely denote capabilities of their products"是原因状语从句。

Module 3　Consolidation Exercise 巩固练习

K15.PDF

I. Answer the following questions according to the passage

1. How does CSMA/CD improve CSMA performance?

2. What must a station do when its frame is ready for transmission?

3. What must a station do in the same time when it is transmitting data?

4. How can collisions be detected?

5. Why modern Ethernet networks do not need to utilize CSMA/CD?

6. How does IEEE 802.5 eliminate collisions?

7. What does IEEE 802 refer to?

8. How is the maximum jam-time calculated?

II. Correct the mistakes in the following sentences

1. The procedure of CDAM/CD can be described like this:

（1）Is my frame ready for transmission? If not, it goes on to the next point. Is medium idle? If yes, wait until it becomes ready.

（2）Start transmitting or monitor for collision during transmission.

（3）Did a collision occur? If so, go to start transmitting.

（4）Reset retransmission counters and end frame transmission.

2. The following procedure is used to resolve a detected collision. The procedure is complete when retransmission is initiated or the retransmission is aborted due to numerous collisions.

（1）Continue transmission (with a frame header/data/CRC signal instead of jam) until maximum packet time is reached to ensure that all receivers detect the collision.

（2）Decrease retransmission counter.

（3）Was the minimum number of transmission attempts reached? If so, abort transmission.

（4）Calculate and wait random back off period based on number of collisions.

（5）Re-enter main procedure at stage.

III. Translate the following sentences into Chinese

1. CSMA/CD uses a carrier sensing scheme in which a transmitting data station detects other signals while transmitting a frame, stops transmitting that frame, transmits a jam signal, and then waits for a random time interval before trying to resend the frame.

2. Methods for collision detection are media dependent, but on an electrical bus such as 10BASE5 or 10BASE2, collisions can be detected by comparing transmitted data with received data or by recognizing a higher than normal signal amplitude on the bus.

3. Importantly, late collisions are not re-sent by the NIC unlike collisions occurring before the first 64 octets; it is left for the upper layers of the protocol stack to determine that there was loss of data.

4. Modern Ethernet networks, built with switches and full-duplex connections, no longer need to utilize CSMA/CD because each Ethernet segment, or collision domain, is now isolated.

5. IEEE Std 802.3, which defines all Ethernet variants, for historical reasons still bears the title "Carrier Sense Multiple Access with Collision Detection (CSMA/CD) access method and Physical Layer specifications".

6. Token Ring Local Area Network (LAN) technology, which uses a special three-byte frame called a "token" that travels around a logical "ring" of workstations or servers, is a communications protocol for Local Area Networks.

IV. Describe the workflow of CSAM/CD using a flow chart

Lesson 16   IP and DNS

Module 1   Text Study 课文学习

Basic Training 基本训练

Text 1   根据语音和视频完成以下任务。

Task 1-0 听录音，记录关键词，理解课文大意。

L16-1-1.MP3

L16-1-2.MP3

L16-1-3.MP3

L16-1.MP4

Task 1-1 辨析词性和词义。

① identity_____   identify_____   identifier_____

② 有"数"含义的词汇有 number、figure 和 digit 等，_____侧重于数字的形状，更适合用来表示个位数 0、1 等，而两位以上的数就不适合了，作动词"计算"时的同义词是_____。

③ _____本义是手指、脚趾，所以同 figure 一样指个位数，在应用中更多用作"位数"，

例如，1 位数_____、2 位数_____、3 位数_____，数码相机用形容词_____（数码的）。

④ _____是纯数学上的"数值"，无论几位数，numeral 的词性是_____，词义是_____。

⑤ numerical/numeric 的词义是_____。

⑥ middle 和 medium 都有"中"的意思，比较它们的不同含义。

_____

Task 1-2 解释术语。

① DNS_____

② DHCP_____

Task 1-3 翻译。

子网掩码_____	单播_____	组播_____
广播_____	A 类 IP 地址_____	属性标签_____
大型网络_____	中型网络_____	小型网络_____
二进制的_____	八进制的_____	十进制的_____
十六进制的_____	二进制_____	八进制_____
十进制_____	十六进制_____	

十进制数字系统由 0 到 9 的数字组成。

_____

IP 地址有 4 类，分别是 0 类、1 类、2 类和 3 类，2 类和 3 类是常用的。

_____

Task 1-4 回答问题或翻译。

① "从 0 开始（含 0）"可以用哪些介词？_____

② "The range of the IP addresses in the class A is between 1 to 126." 含边界"1"和"126"吗？

_____

③ "从 1 到 126（含 1 和 126）"还可以如何表达？

_____

④ from 1 to 100 和 between 1 to 100 有区别吗？

_____

⑤ 16 million_____；4.3 billion_____。

课文

IPv4

IP address is a unique identifier of a computer on TCP/IP networks and on the Internet. Every computer requires a unique IP address to be a part of the Internet and the IP address is provided by the Internet service providers. Every IP address consists of the 32 bits and a binary system of 0s and 1s. The binary number system consists of only two types of digits 0 and 1. It is easier for us to remember the decimal numbers rather than the binary number system such as 011001101. On a same network segment, all the IP address share the same network address.

TCP/IP protocols are installed by default with the Windows-based operating systems. After the TCP/IP protocols are successfully installed, you need to configure them through the Properties Tab of the Local Area Connection.

There are five classes of the IP addresses such as A, B, C, D and E and only 3 classes are in use. Class D IP addresses are reserved for the multicast group and cannot be assigned to hosts and the E class IP addresses are the experimental addresses and cannot be assigned to the people. Every IP address consists of 4 octets and 32 bits. Every participating host and the devices on a network such as servers, routers, switches, DNS (Domain Name System), DHCP (Dynamic Host Configuration Protocol), gateway, Web server, Internet fax server and printer have their own unique addresses within the scope of the network. The detailed information of Class A, Class B and Class C is Shown in Table 3-1.

Table 3-1　The Detailed Information of Three Classes

Class	Leading bits	Size of network number bit field	Size of rest bit field	Number of networks	Addresses per network	Start address	End address
A	0	8	24	$128\ (2^7)$	$16,777,216\ (2^{24})$	0.0.0.0	127.255.255.255
B	10	16	16	$16,384\ (2^{14})$	$65,536\ (2^{16})$	128.0.0.0	191.255.255.255
C	110	24	8	$2,097,152\ (2^{21})$	$256\ (2^8)$	192.0.0.0	223.255.255.25

Class A: The binary address for the Class A starts with 0. The range of the IP addresses in the class A is between 1 to 126 and the default subnet mask of the class A is 255.0.0.0. Class A supports 16 million hosts on each of 125 networks. An example of the class A is 10.10.1.1. Class A is used for the large networks with many network devices.

Class B: The binary address for the class B starts with 10. The range of the IP address in the class B is between 128 to 191 and the default subnet mask for the class B is 255.255.0.0. Class B supports 65,000 hosts on each of 16,000 networks. An example of the class B address is 150.10.10.10. Class B addresses scheme is used for the medium-sized networks.

Class C: The binary address for the class C starts with 110. The range of the IP addresses in the class C is between 192 to 223 and the default subnet mask for the class C is 255.255.255.0. Class C supports 254 hosts on each of 2 million networks. An example of the Class C IP address is 210.100.100.50. Class C is used for the small networks with less than 256 devices and nodes in a network.

Class D: The binary address for the class D starts with 1110 and the IP addresses range can be between 224 and 239. An example of the class D IP address is 230.50.100.1.

Class E: The binary address can start with 11110 and the decimal can be anywhere from 240 to 255. An example of the class E IP address is 245.101.10.10.

Text 2　根据语音和视频完成以下任务。

Task 2-0 听录音，记录关键词，理解课文大意。

L16-2.MP3

_____

_____

_____

Task 2-1 翻译。

① the most recent version＿＿＿＿＿＿；the latest version＿＿＿＿＿；the last version＿＿＿＿＿。

② IPv6 addresses are represented as eight groups of four hexadecimal digits with the groups being separated by colons.

L16-2.MP4

Task 2-2 填空、翻译及词汇辨析。

assign、distribute、divide 和 allocate 在教材中都直接或间接出现过，都有"分配"的意思。

① ＿＿＿＿＿是"上对下的指派"，不一定公平、合理。

翻译：老师给每个儿童布置的作业都不相同。

---

② ＿＿＿＿＿是"分配下去"。

翻译：这个机构向地震灾民分发了食品。

---

③ ＿＿＿＿＿侧重"分配、分开"的动作本身，默认是公平合理的。

翻译：我们共同分担这项工作。

---

④ ＿＿＿＿＿主要指金钱、财产、权力或领土等的分配，侧重于分配的比例和专门用途。

翻译：他们打算给成人学生提供更多的名额。

Task 2-3 比较 limit 和 restrict 的区别，除动词"限制"的意思外，它们在词义、词性上有什么不同，举例说明。

课文

IPv6

Internet Protocol version 6 (IPv6) is the most recent version of the Internet Protocol (IP), the communications protocol that provides an identification and location system for computers on networks and routes traffic across the Internet. IPv6 was developed by the Internet Engineering Task Force (IETF) to deal with the long-anticipated problem of IPv4 address exhaustion. IPv6 is intended to replace IPv4.

Every device on the Internet is assigned an IP address for identification and location definition. With the rapid growth of the Internet after commercialization in the 1990s, it became evident that far more addresses than the IPv4 address space available were necessary to connect new devices in the future. By 1998, the Internet Engineering Task Force (IETF) had formalized the successor protocol. IPv6 uses a 128-bit address, theoretically allowing 2^{128}, or approximately 3.4×10^{38} addresses. The actual number is slightly smaller, as multiple ranges are reserved for special use or completely excluded from use. The total number of possible IPv6 addresses is more than 7.9×10^{28} times as many as IPv4, which uses 32-bit addresses and provides approximately 4.3 billion addresses. The two protocols are not designed to be interoperable, complicating the transition to IPv6. However, several IPv6 transition mechanisms have been devised to permit communication between IPv4 and IPv6 hosts.

IPv6 provides other technical benefits in addition to a larger addressing space. In particular, it permits hierarchical address allocation methods that facilitate route aggregation across the Internet, and thus limits the expansion of routing tables. The use of multicast addressing is expanded and simplified, and provides additional optimization for the delivery of services. Device mobility, security, and configuration aspects have been considered in the design of the protocol.

IPv6 addresses are represented as eight groups of four hexadecimal digits with the groups being separated by colons, for example 2001:0db8:0000:0042:0000:8a2e:0370:7334, but methods to abbreviate this full notation exist.

Advanced Training 进阶训练

T16.PDF

Task 1 翻译。

① A specific node's domain name is the list of the labels in the path from the node being named to the DNS Tree root.

_____

② 绝对域名（完全限定域名）_____

③ 资源记录可以被客户端检索_____

Task 2 用英文回答问题。

如何提高域名解析的效率？

_____

_____

Task 3 解释。

DHCP_____

Task 4 用英文回答问题。

动态主机配置协议有哪些功能？

_____

_____

_____

DNS

The DNS is an IETF-standard name service. The DNS service enables client computers on your network to register and resolve DNS domain names. These names are used to find and access resources offered by other computers on your network or other networks, such as the Internet. The following are the three main components of DNS:

● Domain name space and associated Resource Records (RRs)—A distributed database of name-related information.

● DNS Name Servers—Servers that hold the domain name space and RRs, and that answer queries from DNS clients.

● DNS Resolvers—The facility within a DNS client that contacts DNS name servers and issues name queries to obtain resource record information.

The domain name space is a hierarchical, tree-structured name space, starting at an unnamed root used for all DNS operations. In the DNS name space, each node and leaf in the domain name space tree represents a named domain. Each domain can have additional child domains. Figure 3-4 illustrates the structure of Internet domain name space.

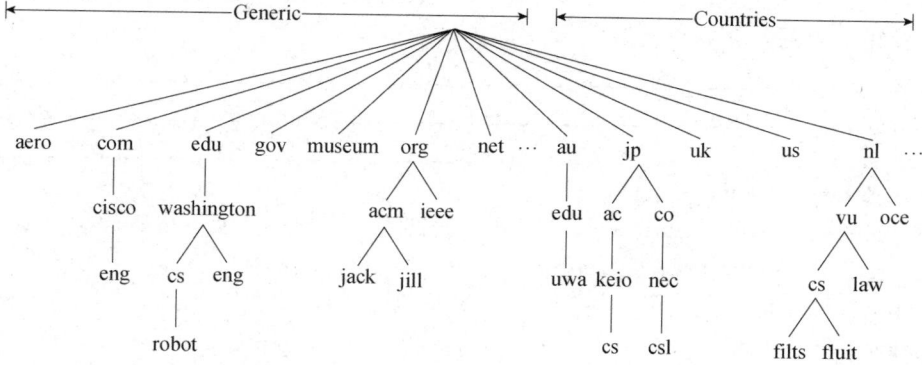

Figure 3-4   Domain Name Space for the Internet

Each node in the DNS tree, as Figure3-4 illustrates, has a separate name, referred to in RFC 1034 as a label. Each DNS label can be from 1 through 63 characters in length, with the root domain having a length of zero characters.

A specific node's domain name is the list of the labels in the path from the node being named to the DNS Tree root. DNS convention is that the labels that compose a domain name are read left to right—from the most specific to the root, for example, www.kapo**.com. This full name is also known as the Fully Qualified Domain Name (FQDN).

A resource record is a record containing information relating to a domain that the DNS database can hold and that a DNS client can retrieve and use. For example, the host RR for a specific domain holds the IP address of that domain (host); a DNS client will use this RR to obtain the IP address for the domain.

Each DNS server contains the RRs relating to those portions of the DNS namespace for which it's authoritative (or for which it can answer queries sent by a host). When a DNS server is authoritative for a portion of the DNS name space, those systems' administrators are responsible for ensuring that the information about that DNS name space portion is correct. To increase efficiency, a given DNS server can cache the RRs relating to a domain in any part of the domain tree.

Dynamic Host Configuration Protocol

The Dynamic Host Configuration Protocol (DHCP) is an auto configuration protocol used on IP networks. Computers that are connected to IP networks must be configured before they can communicate with other computers on the network. DHCP allows a computer to be configured automatically and eliminates the need for intervention by a network administrator. It also provides a central database for keeping track of computers that have been connected to the network. This prevents two computers from accidentally being configured with the same IP address.

In the absence of DHCP, hosts may be manually configured with an IP address. Alternatively IPv6 hosts may use stateless address auto-configuration to generate an IP address. IPv4 hosts may use link-local addressing to achieve limited local connectivity.

In addition to IP addresses, DHCP also provides other configuration information, particularly the IP addresses of local caching DNS resolvers. Hosts that do not use DHCP for address configuration may still use it to obtain other configuration information.

Vocabulary Practice 词汇练习

词汇及短语听写，并纠正发音。

C16.MP3

Words & Expressions

exhaustion [ɪɡˈzɔːstʃən]	*n.* 枯竭，用尽；筋疲力尽，疲惫不堪
commercialization [kəˌmɜːʃəlaɪˈzeɪʃn]	*n.* 商业化，商品化
evident [ˈevɪdənt]	*adj.* 明显的，明白的
formalize [ˈfɔːməlaɪz]	*vt.* 使正式，形式化
theoretically [ˌθɪəˈretɪkli]	*adv.* 理论地，理论上地
interoperable [ˌɪntərˈɒpərəbl]	*adj.* （计算机系统或程序）能互相操作的，能共同使用的
transition [trænˈzɪʃn]	*n.* 过渡，转变
hexadecimal [ˌheksəˈdesɪməl]	*n. & adj.* 十六进制（的）
abbreviate [əˈbriːvɪeɪt]	*vt.* 缩略；使简短；缩简；使用缩写词
notation [nəʊˈteɪʃn]	*n.* 记号，标记法
query [ˈkwɪəri]	*n.* 疑问；询问
convention [kənˈvenʃn]	*n.* 惯例；习俗；规矩；会议
qualify [ˈkwɒlɪfaɪ]	*vi. & vt.* （使）具有资格；达标；合格；陈述
authoritative [ɔːˈθɒrɪtətɪv]	*adj.* 权威的；有权力的；当局的；命令式的

Module 2   Study Aid 辅助帮学

Terms & Abbreviations 术语和缩写

IPv6   *abbr.* 互联网协议第 6 版（Internet Protocol version 6），是 IETF（互联网工程任务组）设计的用于替代现行版本 IP（IPv4）的下一代 IP，号称可以为全世界的每一粒沙子编写一个网址。

RRs   *abbr.* 资源记录（Resource Records），是指每个域所包含的与之相关的资源，即 DNS server 上网域名和 IP 地址对应关系的档案记录。

Difficult Sentences Analysis and Translation 难句分析与翻译

1. Every participating host and the devices on a network such as servers, routers, switches, DNS (Domain Name System), DHCP (Dynamic Host Configuration Protocol), gateway, Web server, Internet fax server and printer have their own unique addresses within the scope of the network.

译文：每个参与的主机及网络上的设备，如服务器、路由器、交换机、DNS（域名系统）服务器、DHCP（动态主机配置协议）、网关、Web 服务器、Internet 传真服务器和打印机，在网络范围内都有自己唯一的地址。

2. Internet Protocol version 6 (IPv6) is the most recent version of the Internet Protocol (IP), the communications protocol that provides an identification and location system for computers on networks and routes traffic across the Internet.

译文：互联网协议第 6 版（IPv6）是互联网协议（IP）的最新版本，是为网络上的计算机和通过互联网传输流量的路由器提供标识和定位系统的通信协议。

分析：本句中"the communications protocol"是同位语，补充说明前面的名词短语；"that"引导的定语从句，修饰、限定"the communications protocol"；"provides…for…"意思是"为……提供……"。

3. With the rapid growth of the Internet after commercialization in the 1990s, it became evident that far more addresses than the IPv4 address space available were necessary to connect new devices in the future.

译文：随着互联网在 20 世纪 90 年代商业化以后的迅速发展，显而易见的是，未来连接新设备所需的地址比 IPv4 地址空间可提供的地址要多得多。

分析：本句介词短语"With the rapid growth…in the 1990s"中"with"表示"随着"，后接名词短语；"it became evident that…"中的"it"是形式主语，"that"引导主语从句。

4. The total number of possible IPv6 addresses is more than 7.9×10^{28} times as many as IPv4, which uses 32-bit addresses and provides approximately 4.3 billion addresses.

译文：IPv6 可能有的地址总数超过 IPv4 地址总数的 7.9×10^{28} 倍，而 IPv4 使用 32 位地址，提供大约 43 亿个地址。

分析：本句中的"IPv6（互联网协议第 6 版）"和"IPv4（互联网协议第 4 版）"是术语缩写，在翻译时也可以不译；句中英语倍数表达方式"…times as many as…"，意思为"是……的……倍"；"which uses…and provides…"是非限制性定语从句，补充说明先行词"IPv4"，翻译时可以单独成句。

5. DNS convention is that the labels that compose a domain name are read left to right—from the most specific to the root, for example, www.kapo**.com.

译文：DNS 的惯例是，组成一个域名的标签要从左读到右，即从最具体的读到根部，例如：www.kapo**.com。

分析：本句中的"DNS（域名系统）"是术语缩写，在翻译时也可以不译；句中"that"引导表语从句；在表语从句中，"that compose a domain name"是定语从句，修饰、限定前面的名词"the labels"，破折号后的内容用于补充说明。

6. When a DNS server is authoritative for a portion of the DNS name space, those systems' administrators are responsible for ensuring that the information about that DNS name space portion is correct.

译文：当 DNS 服务器对 DNS 名称空间的一部分具有权威性时，相关的系统管理员负责确保有关 DNS 名称空间部分的信息是正确的。

分析：本句中的"DNS（域名系统）"是术语缩写，在翻译时也可以不译；句中"When a DNS server is authoritative for a portion of the DNS name space"是时间状语从句；动名词短语"ensuring that …"是短语的"are responsible for"的介词宾语；宾语从句"that the information about that DNS name space portion is correct"作动词"ensure"的宾语。

Module 3   Consolidation Exercise 巩固练习

K16.PDF

I. Answer the following questions according to the passage

1. Is IP address very important for a computer on the Internet? Why?

2. Which is easier to remember, the binary number system or the decimal numbers?

3. Which IP addresses are reserved for the multicast group?

4. Why was IPv6 developed by the Internet Engineering Task Force (IETF)?

5. What does a resource record contain?

6. What is DHCP used to do?

7. How is the domain name space structured?

8. What kind of IP address does 245.101.11.11 belong to?

II. Fill in each blank with information from the passage

1. Every computer requires a _____ to be a part of _____ and the IP address is provided by the Internet _____.

2. Every IP address consists of the _____ and a binary system of 0s and 1s. _____ _____ consists of only two types of _____ 0 and 1.

3. TCP/IP protocols are installed by _____ with the _____ operating systems. After the TCP/IP protocols are successfully installed, you need to _____ them through the _____ of the Local Area Connection.

4. The _____ address for the _____ starts with 0. The _____ of the IP addresses in the class A is between 1 to 126 and the _____ of the class A is 255.0.0.0.

5. In particular, IPv6 permits _____ address allocation methods that facilitate route _____

across the Internet, and thus limits the expansion of _____.

6. In the DNS _____ space, each _____ in the domain name space tree represents a _____ domain. Each domain can have additional _____.

7. When a DNS server is _____ for a _____ of the DNS name space, those systems' _____are responsible for ensuring that the information about that DNS name space portion is _____.

8. In addition to IP addresses, DHCP also provides other _____ information, particularly the IP addresses of local _____ DNS resolvers. Hosts that do not use DHCP for address configuration may still use it to _____ other configuration information.

III. Explain the following terms in English

1. IP address

2. Resource Records

3. RFC

4. IP Addresses Classes

5. DNS

6. DHCP

Lesson 17　　Routing Protocols

Module 1　Text Study 课文学习

Basic Training 基本训练

Text 1　根据语音和视频完成以下任务。

L17-1-1.MP3

Task 1-0 听录音，记录关键词，理解课文大意。

_____

_____

L17-1-2.MP3

_____

_____

Task 1-1　翻译及术语解释。

① IGP_____

② EGP_____

③ BGP_____

④ IS-IS_____

⑤ RIP_____

⑥ OSPF_____

⑦ RFC_____

⑧ 链路状态路由协议_____

⑨ 距离矢量路由协议_____

Task 1-2　翻译。

路由环_____　路由聚合_____　自治系统_____

Task 1-3　词汇辨析。

direct 和 immediate_____

_____

Task 1-4　用英语陈述下列短语。

① 路由协议的作用。

_____

_____

② 路由协议分类的依据。

_____

_____

L17-1.MP4

课文

Routing Protocols and the Classifications

A routing protocol specifies how routers communicate with each other, disseminating information that enables them to select routes between any two nodes on a computer network. Routing algorithms determine the specific choice of route. Each router has a priori knowledge only of networks attached to it directly. A routing protocol shares this information first among immediate neighbors, and then throughout the network. This way, routers gain knowledge of the topology of the network.

Many routing protocols are defined in documents called RFCs. Some versions of the Open System Interconnection (OSI) networking model distinguish routing protocols in a special sublayer of the Network Layer (Layer 3). The specific characteristics of routing protocols include the manner in which they avoid routing loops, the manner in which they select preferred routes, using information about hop costs, the time they require to reach routing convergence, their scalability, and other factors.

Although there are many types of routing protocols (Figure 3-5), three major classes are in widespread use on IP networks:

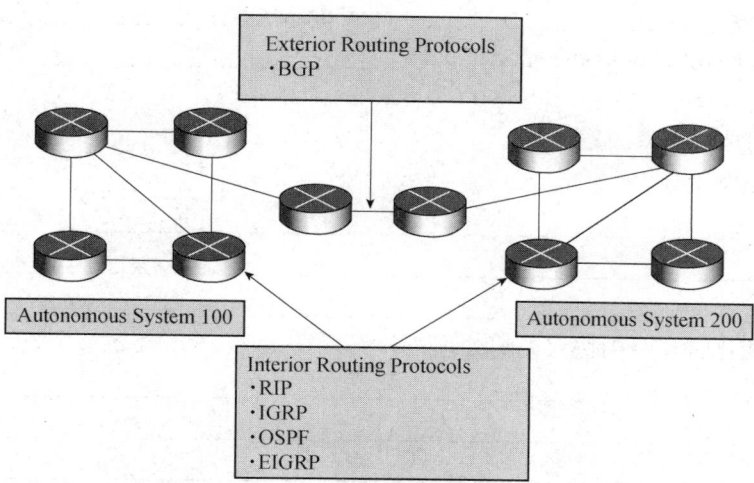

Figure 3-5   Types of Routing Protocols

Interior Gateway Protocols (IGP) type 1, link-state routing protocols, such as OSPF and IS-IS.

Interior Gateway Protocols type 2, distance-vector routing protocols, such as Routing Information Protocol (RIP), RIPv2, IGRP.

Exterior Gateway Protocols, path-vector routing protocols, such as Border Gateway Protocol (BGP).

Exterior Gateway Protocols are routing protocols used on the Internet for exchanging routing information between Autonomous Systems while Interior Gateway Protocols within each of them.

Please notice: the term "Exterior Gateway Protocol (EGP)" has two meanings. It could mean a category of protocols used to exchange routing information between Autonomous Systems. It could also mean a specific RFC-described protocol. It is the same with Interior Gateway Protocol.

Another way of classifying routing protocols is by the type of algorithm they use to determine the best path to a destination network. Routing protocols can be classified as distance vector, link-state, or path vector.

Text 2　根据语音和视频完成以下任务。

Task 2-0 听录音，记录关键词，理解课文大意。　　　　L17-2-1.MP3

_____

_____

_____

L17-2-2.MP3

_____

Task 2-1 分别用 disseminate 和 advertise 造句，体会词义。　　L17-2.MP4

_____

_____

Task 2-2 补充下列句子。

① Distance is _____ and direction is _____.

② The router will choose the shorter path _____ the next-hop router.

Task 2-3 用英语回答问题。

① 怎样避免路由环？

_____

_____

② 当路由变化时，路由协议如何处理？

_____

_____

③ 路由发现的过程是怎样的？

_____

_____

Task 2-4 翻译。

① As routes are learned from one neighbor, that information is passed on to other neighbors with an increase in the routing metric.

_____

② "带毒性逆转的水平分割"的意思是：最好用一个度量值说明该路由不可达，以此毒化该路由，从而明确指出该路由器没有到这个网络的路由。

_____

_____

_____

课文

How Distance Vector Routing Protocols Work

Distance vector means that routes are advertised as vectors of distance and direction. Distance is defined in terms of a metric such as hop count and direction is simply the next-hop router or exit interface. Distance vector routing protocols include: RIPv1, RIPv2, IGRP and EIGRP.

Routers that use distance vector routing protocols determine the best path to remote networks based on the information they learn from their neighbors. If Router X learns of two paths to the same network, one through Router Y at 7 hops, and another route through Router Z at 10 hops, the router will choose the shorter path using Router Y as the next-hop router. Router X has no knowledge of what the network looks like beyond Routers Y and Z, and can only make its best path decision based on the information sent to it by these two routers. Distance vector routing protocols do not have a map of the topology as link-state routing protocols do.

Network discovery is an important process of any routing protocol. Some distance vector routing protocols such as RIP go through a step-by-step process of learning and sharing routing information with their neighbors. As routes are learned from one neighbor，that information is passed on to other neighbors with an increase in the routing metric.

Routing protocols also need to maintain their routing tables to keep them current and accurate. RIP exchanges routing table information with its neighbors every 30 seconds. EIGRP, another distance vector routing protocol, does not send these periodic updates and only sends a "bounded" update when there is a change in the topology and only sends to those routers that need the information.

RIP also uses timers to determine when a neighboring router is no longer available, or when some of the routers may not have current routing information. This is typically because the network has not yet converged due to a recent change in the topology. Distance vector routing protocols also use triggered updates to help speed up convergence time.

One disadvantage of distance vector routing protocols is the potential for routing loops. Routing loops can occur when the network is in a diverged state. Distance vector routing protocols use hold-down timers to prevent the router from using another route to a recently down network until all of the routers have had enough time to learn about this change in the topology.

Split horizon and split horizon with poison reverse are also used by routers to help prevent routing loops. The split horizon rule states that a router should never advertise a route through the interface from which it learned that route. Split horizon with poison reverse means that it is better to explicitly state that this router does not have a route to this network by poisoning the route with a metric stating that the route is unreachable.

Advanced Training 进阶训练

T17.PDF

Task 1 解释术语。

① LSP_____

② LSDB_____

Task 2 用英文解释为什么链路状态路由协议不会出现路由环路。

_____

_____

_____

Task 3 翻译。

① Each router floods the LSP to all neighbors, who then store all LSPs received in a database.

_____

② Each router will have identical LSDBs.

_____

Task 4 填充句子。

① Each router _____ a Link-State Database.

② Each router uses the database to _____a complete map of the topology.

③ When a link is_____, _____or_____, the router will flood the new LSP to all other routers.

④ When a router _____ the new LSP, it will _____ its LSDB, _____ the SPF algorithm, _____ a new SPF tree, and _____ its routing table.

⑤ Link-state routing protocols do require more _____ and _____.

Link-State Routing Protocols

Distance vector routing protocols are sometime referred to as "routing by rumor", although this can be somewhat of a misnomer. Distance vector routing protocols are very popular with many network administrators as they are typically easily understood and simple to be implemented. This does not necessarily mean link-state routing protocols are any more complicated or difficult to configure. Link-state routing protocols are as easy to understand and configure as distance vector routing protocols.

Link-state routing protocols are also known as Shortest Path First (SPF) protocols and are built around Edsger Dijkstra's Shortest Path First algorithm.

There are two link-state routing protocols for IP: OSPF (Open Shortest Path First) and IS-IS (Intermediate System-to-Intermediate System).

The link-state process can be summarized as follows:

- Each router learns about its own directly connected networks.
- Each router is responsible for "saying hello" to its neighbors on directly connected networks.
- Each router builds a Link-State Packet (LSP) containing the state of each directly connected link.
- Each router floods the LSP to all neighbors, who then store all LSPs received in a database.
- Each router uses the database to construct a complete map of the topology and computes the best path to each destination network.

A link is an interface on the router. A link-state is the information about that interface including its IP address and subnet mask, the type of network, the cost associated with the link, and any neighbor routers on that link.

Each router determines its own link-states and floods the information to all other routers in the area. As a result, each router builds a Link-State DataBase (LSDB) containing the link-state information from all other routers. Each router will have identical LSDBs. Using the information in the LSDB, each router will run the SPF algorithm. The SPF algorithm will create an SPF tree, with the router at the root of the tree. As each link is connected to other links, the SPF tree is created. Once the SPF tree is completed, the router can determine on its own the best path to each network in the tree. This best path information is then stored in the router's routing table.

Link-state routing protocols build a local topology map of the network that allows each router to determine the best path to a given network. A new LSP is sent only when there is a change in the topology. When a link is added, removed or modified, the router will flood the new LSP to all other routers. When a router receives the new LSP, it will update its LSDB, rerun the SPF algorithm, create a new SPF tree, and update its routing table.

Link-state routing protocols tend to have a faster convergence time than distance vector routing protocols. A notable exception is EIGRP. However, link-state routing protocols do require more memory and processing requirements. This is usually not an issue with today's newer routers.

Vocabulary Practice 词汇练习

C17.MP3

词汇及短语听写，并纠正发音。

Words & Expressions

disseminate [dɪˈsemɪneɪt]	*vt.* 散布，传播	
priori [praɪˈɔːraɪ]	*n. & adj.* 先验（的），通常作为 a priori 使用	
immediate [ɪˈmiːdɪət]	*adj.* 直接的，最接近的；立即的，立刻的	
exterior [ɪkˈstɪərɪə(r)]	*adj.* 外面的，外部的，外表上的，表面的	
loop [luːp]	*n.* 圈，环	
convergence [kənˈvɜːdʒəns]	*n.* 收敛；集合；会聚	
metric [ˈmetrɪk]	*n.* 度量标准　*adj.* 米制的；公制的；用公制测量的	
trigger [ˈtrɪɡə(r)]	*vt.* 触发；扣……的扳机	
periodic [ˌpɪərɪˈɒdɪk]	*adj.* 周期的；定期的；回归的；间歇的	
diverge [daɪˈvɜːdʒ]	*vi.* 分开；叉开；分歧	
explicitly [ɪkˈsplɪsɪtlɪ]	*adv.* 明确地；直言地；清晰地	
rumor [ˈruːmə(r)]	*n.* 传闻；谣言	
misnomer [ˌmɪsˈnəʊmə(r)]	*n.* 使用不当的名字或名称；用词不当	

Module 2　Study Aid 辅助帮学

Terms & Abbreviations 术语和缩写

IGP　　*abbr.* 内部网关协议（Interior Gateway Protocols），是一个自治网络内的网关交换路由信息的协议。

OSPF　　*abbr.* 开放式最短路径优先（Open Shortest Path First），是一种内部网关协议（IGP），用于在单一自治系统（AS）内决策路由，是链路状态路由协议的一种实现。

IS-IS　　*abbr.* 中间系统到中间系统（Intermediate System-to-Intermediate System）路由协议，最初是 ISO 为无连接网络协议（Connection Less Network Protocol，CLNP）设计的一种动态路由协议。

RIP　　*abbr.* 路由信息协议（Routing Information Protocol），一种基于距离矢量的路由协议，以路由跳数作为计数单位，适用于比较小的网络环境。

IGRP　　*abbr.* 内部网关路由协议（Interior Gateway Routing Protocol），使用组合用户配置尺度（延迟、带宽、可靠性和负载等）决定路由。

Autonomous System　　自治系统，是一个有权自主地决定在本系统中应采用何种路由协议的网络区域。

BGP　　*abbr.* 边界网关协议（Border Gateway Protocol），是运行于自治系统的一种路由

协议，是唯一一个用来处理像互联网大小的网络的协议，也是唯一能够妥善处理好不相关路由域间的多路连接的协议。

　　EGP　　*abbr*. 外部网关协议（Exterior Gateway Protocol），是一种在自治系统的相邻两个网关交换路由信息的协议，通常用于在互联网主机间交换路由表信息。

　　EIGRP　　*abbr*. 增强内部网关路由协议（Enhanced Interior Gateway Routing Protocol），是 Cisco 公司的私有协议。

　　LSP　　*abbr*. 链路状态包（Link-State Packet），是各链路之间用于宣告链路和链路状态的数据包，转发不依靠路由计算，一旦有链路断开或有其他路由传来的 LSP，路由就会更新链路状态表，并转发 LSP。

　　LSDB　　*abbr*. 链路状态数据库（Link-State Database），通过路由器间的路由信息交换，自治系统内部可以达到信息同步，即 LSDB 描述的网络拓扑同步。

Difficult Sentences Analysis and Translation 难句分析与翻译

1. A routing protocol specifies how routers communicate with each other, disseminating information that enables them to select routes between any two nodes on a computer network.

译文：路由协议明确规定路由器彼此通信的方式，并传播信息使它们能够在计算机网络上的任意两个节点之间选择路由。

分析：本句中的宾语从句 "how routers communicate with each other" 作动词 "specifies" 的宾语；现在分词短语 "disseminating information" 表示伴随、同时发生的动作；定语从句 "that enables them to select routes…" 修饰、限定其前的名词 "information"。

2. The specific characteristics of routing protocols include the manner in which they avoid routing loops, the manner in which they select preferred routes, using information about hop costs, the time they require to reach routing convergence, their scalability, and other factors.

译文：路由协议的具体特点包括：避免路由环路的方式、选择优先路由的方式、下一跳花费信息的使用、路由收敛所需的时间、它们的可伸缩性及其他因素。

分析：本句中 "in which they avoid routing loops" 是定语从句，修饰、限定 "the manner"；"in which they select preferred routes" 是定语从句，修饰、限定先行词 "the manner"；"they require to reach routing convergence" 是定语从句，修饰、限定先行词 "the time"；动名词短语 "using information about hop costs"、名词短语 "the time they require to reach routing convergence" "their scalability" 和 "other factors" 都是主句动词 "include" 的宾语。

3. If Router X learns of two paths to the same network, one through Router Y at 7 hops, and another route through Router Z at 10 hops, the router will choose the shorter path using Router Y as the next-hop router.

译文：如果路由器 X 学到两条到同一个网络的路径，一条通过 7 跳到路由器 Y，另一条通过 10 跳到路由器 Z，那么路由器将选择较短的路径，即使用路由器 Y 作为下一跳的路由器。

分析：在本句中，"If" 引导条件状语从句，其宾语是 "two paths to the same network"；"one through Router Y at 7 hops, and another route through Router Z at 10 hops" 是同位语，补充说明前面的名词短语 "two paths to the same network"；现在分词短语 "using Router Y as the next-hop router" 是伴随状语，相当于 "and use…"。

4. EIGRP, another distance vector routing protocol, does not send these periodic updates and only sends a "bounded" update when there is a change in the topology and only sends to those routers that need the information.

译文：EIGRP，另一个距离矢量路由协议，不发送这些周期性的更新，而是只在拓扑有变化时发送一个"有限的"更新，而且只发送给那些需要信息的路由器。

分析：本句中的"EIGRP"是术语缩写，在翻译时也可以不译；"another distance vector routing protocol"是同位语，补充说明前面的术语；"when there is a change in the topology"是时间状语从句；"that need the information"是定语从句，修饰、限定"those routers"。

5. Distance vector routing protocols use hold-down timers to prevent the router from using another route to a recently down network until all of the routers have had enough time to learn about this change in the topology.

译文：距离向量路由协议使用抑制定时器来防止路由器使用另一条到最近中断了的网络的路由，直到所有的路由器都有足够的时间来获知拓扑中的这种变化。

分析：本句中动词短语"prevent...from using..."的意思是"防止……使用……"；"until all of the routers have had enough time to learn about this change in the topology"是时间状语从句。

6. Split horizon with poison reverse means that it is better to explicitly state that this router does not have a route to this network by poisoning the route with a metric stating that the route is unreachable.

译文：带毒性逆转的水平分割的意思是，最好用一个度量值说明该路由不可达，以此毒化该路由，从而明确指出该路由器没有到这个网络的路由。

分析：本句中的宾语从句"that it is better to explicitly state that ... network"作动词"means"的宾语；在该宾语从句中，"that this router does not have a route to this network"也是宾语从句，作动词"state"的宾语；介词短语"by poisoning the route with a metric"作方式状语；现在分词短语"stating that the route is unreachable"作定语，相当于"which states..."，修饰、限定前面的名词"a metric"，其中宾语从句"that the route is unreachable"作动词"stating"的宾语。

Module 3   Consolidation Exercise 巩固练习

K17.PDF

I. Give the meaning of the following abbreviated terms both in English and in Chinese

| OSPF | IS-IS | RIP | IGRP |
| BGP | EIGRP | LSP | LSDB |

II. Fill in each blank with appropriate words or expressions according to the passage

1. A routing protocol specifies how routers _____ with each other, _____ information that enables them to select routes between any two _____ on a computer network.

2. Exterior Gateway Protocols are _____ protocols used on the Internet for exchanging routing information between _____, such as Border _____ Protocol (BGP), _____ Vector Routing Protocol.

3. The specific characteristics of routing protocols include the _____ in which they avoid routing _____, the manner in which they select preferred routes, using information about _____ costs, the time they require to reach routing _____, their _____, and other factors.

4. Routing protocols can be classified as distance _____, _____ state, or _____ vector.

5. RIP exchanges routing _____ protocols information with its neighbors every 30 seconds. EIGRP, another _____ vector routing protocol, does not send these _____ updates and only sends a "bounded" update when there is a change in the _____ and only sends to those _____ that need the information.

6. When a _____ is added, removed or _____, the router will _____ the new LSP to all other routers. When a router receives the new _____, it will update its _____, rerun the SPF algorithm, create a new SPF _____, and update its routing _____.

7. Split _____ with poison reverse means that it is better to explicitly state that this router does not have a route to this network by _____ the route with a _____ stating that the route is _____.

III. Translate the following sentences into Chinese

1. A routing protocol shares this information first among immediate neighbors, and then throughout the network.

2. The term "Exterior Gateway Protocol" has two meanings. It could mean a category of protocols used to exchange routing information between Autonomous Systems. It could also mean a specific RFC-described protocol.

3. Routers that use distance vector routing protocols determine the best path to remote networks based on the information they learn from their neighbors.

4. Distance vector routing protocols use hold-down timers to prevent the router from using another route to a recently down network until all of the routers have had enough time to learn about this change in the topology.

5. Split horizon and split horizon with poison reverse are also used by routers to help prevent routing loops.

6. Link-state routing protocols are also known as Shortest Path First protocols and are built around Edsger Dijkstra's Shortest Path First (SPF) algorithm.

7. When a router receives the new LSP, it will update its LSDB, rerun the SPF algorithm, create a new SPF tree, and update its routing table.

IV. Make sentences about network technology including the following words

1. convergence

2. scalability

3. loop

4. periodic

5. exterior

Lesson 18   OSI Reference Model and TCP/IP

Module 1   Text Study 课文学习

Basic Training 基本训练

Text 1   根据语音和视频完成以下任务。

Task 1-0 听录音，记录关键词，理解课文大意。

L18-1-1.MP3

L18-1-2.MP3

_____
_____
_____

L18-1.MP4

Task 1-1 用英语写出 OSI 模型中七层的名称。

Task 1-2 比较下列词的词性和词义。
① descript、illustrate、explain。

② above、below、top、bottom、up、down、upper、under。

③ dividing、separation、break。

④ part、component。

⑤ accelerate、facilitate。

⑥ perform、carry on。

Task 1-3 翻译。
电压振幅＿＿＿＿＿＿＿＿＿＿＿＿＿＿　比特持续时间＿＿＿＿＿＿＿＿＿＿＿＿＿
差错控制＿＿＿＿＿＿＿＿＿＿＿＿＿＿＿　差错恢复＿＿＿＿＿＿＿＿＿＿＿＿＿＿＿
透明地传输＿＿＿＿＿＿＿＿＿＿＿＿＿＿

Task 1-4 用英语解释术语。
① OSI＿＿＿＿＿＿＿＿＿＿＿＿＿＿＿＿＿＿＿＿＿＿＿＿＿＿＿＿＿＿＿＿＿＿＿＿
② laying＿＿＿＿＿＿＿＿＿＿＿＿＿＿＿＿＿＿＿＿＿＿＿＿＿＿＿＿＿＿＿＿＿＿＿

③ layer＿＿＿＿＿＿＿＿＿＿＿＿＿＿＿＿＿＿＿＿＿＿＿＿＿＿＿＿＿＿＿＿＿＿＿

Task 1-5 造句。
① 分别用 concern 的名词和动词造句。

② 分别用 be short for 和 short for 造句。

Task 1-6 以"over/across/through the physical link"为例，比较介词的不同含义。

课文

OSI Reference Model

The Open Systems Interconnection reference model (OSI reference model or OSI model for short), released in 1984, is a layered, abstract description for communications and computer network protocol design. The OSI reference model is the primary model for network communications. A primary objective of the OSI reference model is to accelerate the development of future networking products. Although there are other models in existence, most network vendors today relate their products to the OSI reference model, especially when they want to educate users to use their products. They consider it the best tool available to teach people about sending and receiving data on a network.

The OSI reference model includes seven layers. The layers, described below and shown in Figure 3-6 are, from bottom to top: Physical Layer, Data Link Layer, Network Layer, Transport Layer, Session Layer, Presentation Layer and Application Layer. Each layer illustrates a particular network function. This separation of networking functions is called layering. A layer is a collection of related functions, which provides services to the layer above it and receives service from the layer below it. Dividing the network into these seven layers provides the following advantages:

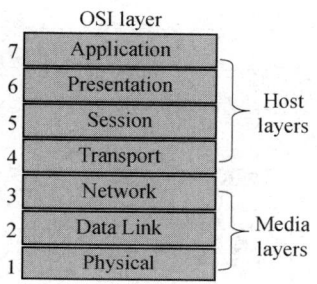

Figure 3-6　OSI reference model layers

- It breaks network communication into smaller, simpler parts that are easier to develop.
- It facilitates standardization of network components to allow multiple vendors' development and support.
- It allows different types of network hardware and software to communicate with each other.
- It prevents changes in one layer from affecting the other layers, so that they can develop more quickly.
- It breaks network communication into smaller parts to make learning it easier to understand.

Physical Layer — It is concerned with transmission of unstructured bit stream over the physical link, and it invokes such parameters as signal voltage swing and bit duration. It defines the mechanical, electrical, procedural and functional characteristics to establish, maintain and deactivate the physical link between end systems.

Data Link Layer — It provides reliable transfer of data across the physical link. It sends blocks of data (frames) with the necessary synchronization, error control and flow control.

Network Layer — It provides upper layers with independence from the data transmission and switching technologies used to connect systems. It is responsible for establishing, maintaining and terminating connections.

Transport Layer — It provides reliable, transparent transfer of data between end points. It is responsible for end-to-end error recovery and flow control.

Session Layer — It provides the control structure for communication between applications. It establishes, manages and terminates connections (sessions) between cooperating applications.

Presentation Layer — It performs transformations on data to provide a standardized application interface and to provide common communications services. It provides services such as encryption, text compression and reformatting.

Application Layer — Provides services such as FTP, HTTP, Telnet, etc. to the users.

Text 2　根据语音和视频完成以下任务。

Task 2-0 听录音，记录关键词，理解课文大意。

L18-2-1.MP3

_____

_____

_____

L18-2-2.MP3

_____

_____

L18-2.MP4

Task 2-1 根据课文解释术语（中文或英文）。
① TCP_____
② UDP_____
③ 数据报_____
④ IP_____

Task 2-2 翻译并比较 equivalent 和 equal。
① Provides services that are roughly equivalent to the OSI Network layer.

_____

② Provides services that are roughly equal to the OSI Network layer.

_____

Task 2-3 翻译。
① It invokes such parameters as signal voltage swing and bit duration.

_____

② This includes all of the processes that involve user interaction.

_____

Task 2-4 用英语描述 TCP/IP 模型的层次结构。

_____

Task 2-5 分别用 underlying、under、below 和 lower 表达"网络访问层协议必须知道底层网络的细节"，体会它们的区别。

_____

Task 2-6 表示数据和信息单位的词，除了 bit、byte、packet 外还有哪些？将其从小到大进行排列。

_____

TCP/IP Protocol Hierarchy

As is shown in Figure 3-7, the TCP/IP model has four layers from bottom to top: the Network Access Layer, the Internet Layer, the Transport Layer, and the Application Layer.

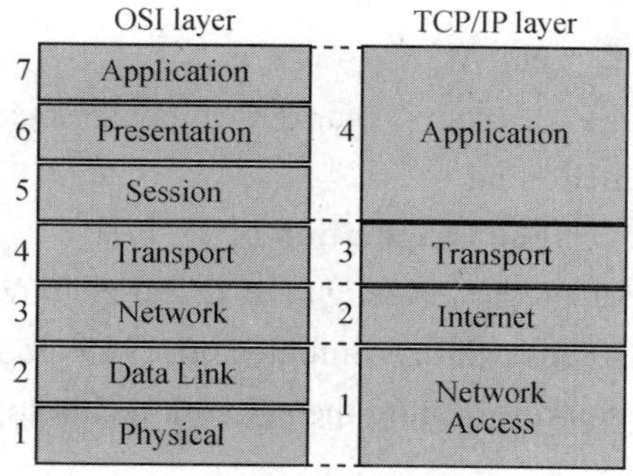

Figure 3-7   OSI and TCP/IP

Network Access Layer — The lowest layer of the TCP/IP protocol hierarchy. It defines how to use the network to transmit an IP datagram. Unlike higher-level protocols, Network Access Layer protocols must know the details of the underlying network (its packet structure, addressing, etc.) to correctly format the data being transmitted to comply with the network constraints. The TCP/IP Network Access Layer can encompass the functions of two lower layers of the OSI reference model (Physical Layer and Data Link Layer). As new hardware technologies appear, new Network Access protocols must be developed so that TCP/IP networks can use the new hardware. Consequently, there are many access protocols — one for each physical network standard.

Access protocol is a set of rules that defines how the hosts access the shared medium. Access protocol has to be simple, rational and fair for all the hosts. Functions performed at this level include encapsulation of IP datagram into the frames transmitted by the network, and mapping of IP addresses to the physical addresses used by the network. One of TCP/IP's strengths is its universal addressing scheme. The IP address must be converted into an address that is appropriate for the physical network over which the datagram is transmitted.

Internet Layer — It provides services that are roughly equivalent to the OSI Network layer. The primary concern of the protocol at this layer is to manage the connections across networks as information is passed from source to destination. The Internet Protocol (IP) is the primary protocol at this layer of the TCP/IP model.

Transport Layer — It is designed to allow peer entities on the source and destination hosts to carry on a conversation, just as in the OSI transport layer. Two end-to-end transport protocols, TCP and UDP, have been defined here.

Application Layer — It includes the OSI Session, Presentation and Application layers as shown in the Figure 3-7. An application is any process that occurs above the Transport Layer. This includes all of the processes that involve user interaction. The application determines the presentation of the data and controls the session. There are numerous application layer protocols in TCP/IP suite, including Simple Mail Transfer Protocol (SMTP) and Post Office Protocol (POP) used for e-mails, Hyper Text Transfer Protocol (HTTP) used for the World-Wide-Web, and File Transfer Protocol (FTP). Most application layer protocols are associated with one or more port numbers.

Advanced Training 进阶训练

T18.PDF

Task 在下列词汇中找出适当的词汇填在画线处。

original   originated   suite   intact   survive   constructed   broken   operation
agency   packet   functioning

The Origination of TCP/IP

TCP/IP _____ out of the investigative research into networking protocols that the United States Department of Defense (DoD) initiated in 1969. In 1968, the Advanced Research Projects _____ (ARPA) began researching the network technology that is called _____ Switching (PS).

The _____ focus of this research was that the network was able to _____ in case of the loss of subnet hardware, with existing conversations not being _____ off. In other words, the DoD wanted the connections to remain _____ as long as the source and destination nodes were _____, even if some of the machines or transmission lines in between were suddenly put out of _____. The network that was initially _____ as a result of this research to provide a communication that could function in wartime, when called ARPANET, gradually became known as the Internet. The TCP/IP protocols played an important role in the development of the Internet. In the early 1980s, the TCP/IP protocols were developed. In 1983, they became standard protocols for ARPANET. Because of the history of the TCP/IP protocol suite, it's often referred to as the DoD protocol _____ or the Internet protocol suite.

Vocabulary Practice 词汇练习

C18.MP3

词汇及短语听写，并纠正发音。

_____

_____

_____

Words & Expressions

swing [swɪŋ]	*n.* 摆动；摇动
procedural [prəˈsiːdʒərəl]	*adj.* 程序的；有关程序的
deactivate [ˌdiːˈæktɪveɪt]	*vt.* 使（仪器等）停止工作；使失活；使（化学过程）灭活化
synchronization [ˌsɪŋkrənaɪˈzeɪʃn]	*n.* 同步；同一时刻；同时性
intact [ɪnˈtækt]	*adj.* 完整无缺的；原封不动的
datagram [ˈdeɪtəˌgræm]	*n.* 数据报（指网络层的数据传输单位）
comply [kəmˈplaɪ]	*vi.* 遵从；依从，顺从
constraint [kənˈstreɪnt]	*n.* 限制；约束；限定
encompass [ɪnˈkʌmpəs]	*vt.* 包含，包括；包围，围绕

rational [ˈræʃnəl]	*adj.* 合理的；理性的；明智的
peer [pɪə(r)]	*n.* 同龄人；同辈；身份（或地位）相同的人；贵族
	vi. 仔细看；端详

Module 2　Study Aid 辅助帮学

Terms & Abbreviations 术语和缩写

DoD　　　*abbr.* 美国国防部（United States Department of Defense），是美国政府下属的一个部门，它的中心是五角大楼。

ARPA　　　*abbr.* 高级研究计划局（Advanced Research Projects Agency），是美国国防部下属的一个行政机构，负责研发用于军事的高新科技。

PS　　　*abbr.* 分组交换（Packet Switching），是指在通信过程中，通信双方以分组为单位，使用存储—转发机制来实现数据交互的通信方式。

UDP　　　*abbr.* 用户数据报协议（User Datagram Protocol），是 OSI 参考模型中一种无连接的传输层协议，提供面向事务的简单、不可靠信息传送服务。

Difficult Sentences Analysis and Translation 难句分析与翻译

1. Although there are other models in existence, most network vendors today relate their products to the OSI reference model, especially when they want to educate users to use their products.

译文：虽然存在其他模型，但是现在大多数网络供应商都将产品与 OSI 参考模型联系起来，特别是当他们想培训用户使用他们的产品时。

分析：本句中 "Although there are other models in existence" 是让步状语从句；动词短语 "relate…to…" 的意思是 "将……与……联系起来"；"especially when they want to educate users to use their products" 是时间状语从句。

2. In other words, the DoD wanted the connections to remain intact as long as the source and destination nodes were functioning, even if some of the machines or transmission lines in between were suddenly put out of operation.

译文：换句话说，美国国防部希望的是，即使有些机器或其中的传输链路突然停止运行，只要源节点和目标节点正常运行，连接就可以保持完好。

分析：本句中 "as long as the source and destination nodes were functioning" 是条件状语从句；"even if some of the machines or transmission lines in between were suddenly put out of operation" 是让步状语从句。

3. The network that was initially constructed as a result of this research to provide a communication that could function in wartime, when called ARPANET, gradually became known as the Internet.

译文：该网络起初作为这项研究的结果而建，是为了提供在战时仍能运作的通信，当时被称为阿帕网，后来逐渐被称为互联网。

分析：本句主干部分为 "The network gradually became known as the Internet"；句中 "that was initially constructed as a result of this research to provide a communication" 是定语从句，修饰、限定 "The network"；"that could function in wartime" 是定语从句，修饰、限定 "communication"；过去分词短语 "when called ARPANET" 作非限制性定语。

4. The IP address must be converted into an address that is appropriate for the physical network over which the datagram is transmitted.

译文：IP 地址必须转换成适合发送数据报的物理网络的地址。

分析：本句中的"IP（网络协议）"是术语缩写，在翻译时也可以不译；句中"that is appropriate for the physical network"是定语从句，修饰、限定"an address"；"over which the datagram is transmitted"是定语从句，修饰、限定"the physical network"。

5. There are numerous application layer protocols in TCP/IP, including Simple Mail Transfer Protocol (SMTP) and Post Office Protocol (POP) used for e-mails, Hyper Text Transfer Protocol (HTTP) used for the World-Wide-Web, and File Transfer Protocol (FTP).

译文：TCP/IP 中有很多应用层协议，包括简单邮件传输协议（SMTP）和用于电子邮件的邮局协议（POP）、万维网中使用的超文本传输协议（HTTP）和文件传输协议（FTP）。

分析：本句中的"TCP/IP（传输控制协议/网络协议）"是术语缩写，在翻译时也可以不译；句中分词短语"including ... (FTP)"作为伴随状语，说明同时发生的动作；过去分词短语"used for e-mails"作定语，修饰、限定其前面的名词短语"Post Office Protocol (POP)"；过去分词短语"used for the World-Wide-Web"作定语，修饰、限定其前面的名词短语"Hyper Text Transfer Protocol (HTTP)"。

Module 3　Consolidation Exercise 巩固练习

K18.PDF

I. Answer the following questions according to the passage

1. What is the primary objective of OSI reference model?

2. How many layers does the OSI reference model include? What are they?

3. What is a layer?

4. What advantages does dividing the network into these seven layers provide?

5. What functions do Physical Layer and Transport Layer have?

6. What layers does the TCP/IP model have?

7. Why did DoD develop TCP/IP?

8. How are the TCP/IP layers related to the OSI reference model?

II. Fill in each blank with information from the passage

1. The Open Systems _____ reference model (OSI reference model or OSI model for short), released in 1984, is a _____, _____ description for communications and computer network protocol design.

2. The OSI reference model includes _____ layers. Each layer illustrates a _____ network function. This separation of networking functions is called _____.

3. Data Link Layer provides _____ transfer of data across the physical link. It sends blocks of data (frames) with the necessary _____, _____ control and _____ control.

4. In the early 1980s, the TCP/IP protocols were _____. In 1983, they became _____ protocols for ARPANET. Because of the history of the TCP/IP protocol suite, it's often _____ as the DoD protocol _____ or the Internet protocol suite.

5. Access protocol is a set of rules that defines how the hosts access the shared medium which has to be _____, _____ and _____ for all the hosts.

6. An application is any process that _____ above the Transport Layer. This includes all of the _____ that involve user interaction. The application _____ the presentation of the data and controls the session.

7. Presentation Layer performs _____ on data to provide a standardized application _____ and to provide common communications services. It provides services such as _____, text compression and reformatting.

III. Translate the following sentences into Chinese

1. Network vendors consider the OSI reference model the best tool available to teach people about sending and receiving data on a network.

2. A layer is a collection of related functions, which provides services to the layer above it and receives service from the layer below it.

3. Physical Layer is concerned with transmission of unstructured bit stream over the physical link, and it invokes such parameters as signal voltage swing and bit duration.

4. The original focus of this research was that the network was able to survive in case of the loss of subnet hardware, with existing conversations not being broken off.

5. Functions performed at this level include encapsulation of IP datagram into the frames transmitted by the network, and mapping of IP addresses to the physical addresses used by the network.

6. There are numerous application layer protocols in TCP/IP, including Simple Mail Transfer Protocol (SMTP) and Post Office Protocol (POP) used for e-mails, Hyper Text Transfer Protocol (HTTP) used for the World-Wide-Web, and File Transfer Protocol (FTP).

7. The network that was initially constructed as a result of this research to provide a communication that could function in wartime, when called ARPANET, gradually became known as the Internet.

IV. Make sentences about IT with the following words

1. datagram

2. peer

3. transparent

4. encapsulation

5. OSI